Guilherme de Freitas Furtado
Lúcia Helena Garófalo Chaves

BIOCHAR FROM CHICKEN LITTER AND NPK

Effect on sunflower growth and yield

ScienciaScripts

Imprint
Any brand names and product names mentioned in this book are subject to trademark, brand or patent protection and are trademarks or registered trademarks of their respective holders. The use of brand names, product names, common names, trade names, product descriptions etc. even without a particular marking in this work is in no way to be construed to mean that such names may be regarded as unrestricted in respect of trademark and brand protection legislation and could thus be used by anyone.

Cover image: www.ingimage.com

This book is a translation from the original published under ISBN 978-620-3-46597-6.

Publisher:
Sciencia Scripts
is a trademark of
Dodo Books Indian Ocean Ltd., member of the OmniScriptum S.R.L Publishing group
str. A.Russo 15, of. 61, Chisinau-2068, Republic of Moldova Europe
Printed at: see last page
ISBN: 978-620-3-65430-1

Guilherme de Freitas Furtado
Lúcia Helena Garófalo Chaves

BIOCHAR FROM CHICKEN LITTER AND NPK

THANKS

To God, for the gift of life and for all the conquests and blessings granted;

To the Federal University of Campina Grande, especially the Coordination of Graduate Studies in Agricultural Engineering, for the opportunity to take this course.

To the Coordination for the Improvement of Higher Education Personnel (CAPES) for granting the scholarship.

INDEX

CHICKEN LITTER BIOCHAR AND NPK
Effect on sunflower growth and yield

SUMMARY

The use of biomass can contribute to the improvement or maintenance of soil properties and, therefore, to the production of sustainable energy and food. Given the importance of sunflower for the semi-arid region, the objective was to evaluate the effect of fertilization with poultry litter biochar in combination with mineral fertilizer, on the growth and production of sunflower cv. Embrapa 122/V2000. The experiment was developed in a protected environment of UAEAg/CTRN/UFCG in pots with 20 dm3 capacity under experimental design entirely randomized, with treatments arranged in a 5 x 4 factorial scheme and four repetitions. The treatments were five doses of mineral fertilization (D1=0, D2=25%, D3=50%, D4=75% and D5=100% of the recommended NPK fertilization (100; 300 and 150 mg kg-1) and four doses of poultry manure, 0; 5% (400g/vase); 10% (800g/vase) and 15% (1100g/vase) calculated based on soil volume. Fertilization with 50% of the NPK recommendation (50:150:75 mg kg-1) promoted the highest results in sunflower growth, phytomass production, production components, and yield when compared to the other treatments at NPK levels. Fertilization with increasing doses of chicken manure biochar promoted linear reduction of plant height (PA), stem diameter (CD), leaf number (LE) and leaf area (LA); however, there was no significant effect on phytomass production. The chapter diameter (DCap), number of achenes per chapter (NAC) and number of viable achenes (NAV), had decreases of 14.75%, 35.66% and 40.03% with the increase of biochar doses. The combination of 50:150:75 mg kg-1 of NPK and 400 g/vessel of biochar promoted the highest production of achenes per plant (23.91 g/plant).

Palavras-c have: biomass, fertilization, *Helianthus annuus* L. productivity

INTRODUCTION

The world population is expected to increase to 9.5 billion by 2050 (U.N., 2013), naturally promoting increased demand for food, highlighting the need to expand agricultural boundaries; however, the use of technologies that improve agricultural production with less environmental impact has become the challenge of the new millennium. In this sense, the application of biochar for agriculture has been proposed as an attractive approach to mitigate greenhouse gas emissions and improve crop productivity (LEHMANN et al., 2006).

Biochar is an organic compound produced through the process of fast or slow pyrolysis, with the purpose of application to the soil, improving its physical-chemical characteristics (LEHMANN et al., 2006). Pyrolysis is defined as the thermal degradation of biomass in the absence or low concentrations of $_{O2}$ to produce condensable vapors, gases and charcoal. Biochar can be produced from different biomass sources, such as woody materials, agricultural residues, (IOANNIDOU & ZABANIOTOU, 2007), green waste (CHAN et al., 2007) and manures (LIMA et al., 2008). However, according to Comer et al. (2012), the production of biochar from animal waste has higher nutritional quality compared to biochar produced from plant waste because of higher nitrogen levels.

Chicken litter is of special interest for the production of biochar in Brazil because of the high production generated per year that, according to Corrêa & Miele (2011) is around 6.8 million m3. According to Santos & Grangeiro (2013) fertilization with chicken litter is a low-cost and economically feasible technique for family farmers in Paraíba, and can contribute significantly to promote greater productivity by ensuring the establishment of this crop within the production system of small farmers.

Lehmann et al. (2006) point out that biochar can be applied for soil correction, increases productivity and improves its chemical properties. Petter et al. (2012) found that in the first year of application eucalyptus biochar positively affected soil fertility, total organic carbon, Ca, P, and pH at a depth of 0-10 cm, and was the only factor with a significant effect on the productivity of upland rice. It is observed, however, greater efficiency of the application of biochar when it is associated with mineral fertilization because of its low availability of nutrients (LEHMANN et al., 2003; STEINER et al., 2007).

Sunflower (*Helianthus annuus* L.) is an oleaginous plant that has been attracting the interest of farmers because of its high potential for the production of biofuel. It is a very important crop for the Brazilian semiarid region because of its good adaptability to high temperatures, accounting for about 13% of all vegetable oil produced in the world, with an increase in the cultivated area in recent years (NOBRE et al., 2010).

Sunflower has desirable characteristics from an agronomic point of view, such as a short cycle, high quality and good oil yield (MORAIS et al., 2011). In addition, its grains, the remains of the crop and the by-products generated in the extraction of oil can be used as animal feed (NOBRE et al., 2011). Its high capacity for nutrient cycling provided by absorption at greater depths and its reduced rate of nutrient export have favored the interest of producers in the crop (CASTRO & OLIVEIRA, 2005).

The Brazilian sunflower production in the 2012/2013 harvest was 110,000 tons with the state of Mato Grosso being the largest producer, with 84,700 tons, which corresponds to 64.9% of national production and an average productivity of 1671 kg ha-1. In the Northeast, the states of Ceará and Bahia stand out, with a planted area of 0.2 and 0.3 thousand ha respectively and an average productivity of 456 and 400 kg ha-1 (CONAB, 2013); however, the ability to express maximum productivity of achenes is related, in part, to adequate mineral nutrition of the plant (AQUINO et al., 2013).

Existing information on the use of biochar in production systems is still scarce, with no studies focused on its use in sunflower cultivation. Recently, the International Biochar Initiative (IBI) has carried out studies aiming to provide standards for the use of biochar and to evaluate its soil properties. Such studies may enable the production of biochar with the desired properties for specific purposes; with this, its characterization is necessary to confirm its real agronomic and environmental effects and its economic viability.

Accordingly, this research aims to evaluate the effect of poultry litter biochar in combination with mineral fertilizer on sunflower cv. Embrapa 122/V2000.

REVIEW OF LITERATURE

2.1 Sunflower cultivation

2.1.1 General aspects

Sunflower, a crop originally from the American continent has Mexico as its center of origin, and is widely cultivated throughout the world. It adapts to different environmental conditions and can be cultivated in Brazil, from the state of Rio Grande do Sul to the state of Roraima (CASTRO & FARIAS, 2005). In Europe, the sunflower was introduced at the end of the 16th century as an ornamental plant, spreading through Spain, Italy, France, Belgium, Holland, Sweden, Germany and England. It started to be used as an oleaginous crop from the 18th century on (DALL AGNOL et al., 2005). This crop was introduced in South America in Argentina in the 19th century by Russian Jewish immigrants, and then expanded to Brazil, Uruguay, Chile, Paraguay and Bolivia (DALL AGNOL et al., 2005). In Brazil, the sunflower was introduced by European colonizers who settled in the south and initially consumed its roasted seeds and made a kind of tea rich in caffeine.

The commercial production of sunflower seeds occurred in the state of São Paulo at the beginning of the 20th century, also standing out as a forage species for the dairy basin and as a honey culture in bird feed. Despite the federal government's initiative at the end of the 1970s, through the Energy Mobilization Program, which stimulated the use of vegetable oils to substitute petroleum derivatives, sunflower did not manage to establish itself as an expressive crop in Brazil (LIRA et al., 2011). It was only in the 1990s that sunflower cultivation became prominent again on the Brazilian agricultural scene, especially in the

Midwest, especially in the states of Goiás and Mato Grosso do Sul (DALL AGNOL et al., 2005).

Currently, sunflower is among the four largest edible vegetable oil producing crops in the world; the world demand for sunflower oil is high, averaging 1.8% per year. According to data from the United States Department of Agriculture (USDA) the world production of sunflower seed was 39.40 million tons with Russia (24.44%), Ukraine (24.12%) and Argentina (8.88%) standing out as the largest producers in the 2011/2012 harvest. Among the producing states, Mato Grosso do Sul stands out with a production of 84.7 thousand tons, which corresponds to 78.35% of national production, with a planted area of 50.7 thousand ha and average productivity of 1,671 kg ha-1 in the 2012/2013 harvest. The Northeast participates in only 0.18% of national production, highlighting the states of Ceará and Bahia with a planted area of 0.2 and 0.3 thousand ha respectively and average productivity of 456 and 400 kg ha-1 (CONAB, 2013).

Besides serving the market of noble edible oils, sunflower has a multitude of uses, such as: confectionery, bird feed, production of silage, bran, cake for animal feed, ornamental production, besides the possibility of exporting grains (LIMA et al., 2013); the leaves can be used as natural herbicides because of its allelopathic potential (SILVA et al., 2009); bread flour, which has been used in the manufacture of mixed bread in mixture with wheat, corn and sorghum flours (SACHS et al., 2005). Sunflower oil has essential fatty acids that contribute to the reduction of bad cholesterol (LDL) levels and the prevention of vascular diseases.

2.1.2 Botany and phenology

Sunflower (*Helianthus* annuus L.) is an annual dicotyledon of the order Asterales, family Asteraceae, subfamily Asteroideae and tribe Heliantheaede (CASTRO & FARIAS, 2005). It has a pivotal root system, an erect stem, generally not branched, with a height between 0.3 and 5.0 m and the number of leaves per plant between 8 and 70 depending on the cultivar and growing conditions (WATANABE, 2007). The inflorescence is a chapter where the grains are developed, which has two types of flowers: the tubular ones, which are fertile and hermaphrodite, and the ligulate ones, which are incomplete flowers, with an ovary and rudimentary calyx and a transformed corolla, similar to a petal, of yellow-orange color (ROSSI, 1998). The seed is actually the fruit, aquenium type, and can be white, black or striped, containing from 38 to 50% oil (SANTOS JUNIOR, 2010). In commercial genotypes the mass of 1000 achenes varies from 30 to 60 g and the most frequent number of achenes can vary between 800 and 1700 per chapter. The vegetative cycle varies between 65 and 155 days depending on the cultivar, the genotype, the date of sowing and the soil and climate conditions of each region and year of cultivation (CASTRO & FARIAS, 2005).

The pollination is entomophilic, with bees of the species *Apis mellifera* L. being the main pollinating agent. According to Ungaro (2000), sunflower pollen has protuberances similar to spines that make it suitable for insect transport; moreover, because of its high weight it moves very little by wind action.

According to Schneiter & Miller (1981) sunflower growth is divided into two developmental stages: Vegetative (V) and Reproductive (R) (Table 1). The vegetative stage is subdivided into two phases, that of emergence and that of development of true leaves (length greater than 4 cm), whose beginning is marked by the emergence of seedlings and the end by the beginning of the appearance of inflorescence, the duration of which depends on the genotype and environmental factors. The reproductive stage is divided into nine phases based on the development of the inflorescence, from its visual appearance to the physiological maturity of the seed.

Table 1 - Schematic description of sunflower development stages, adapted from Schneiter & Miller (1981)

Emergency	Number of True Sheets			Development from Inflorescence				Flowering		Filling from Aquênios		Maturation Physiological
VE	V1	V2	VN	R1	R2	R3	R4	R5	R6	R7	R8	R9
Vegetative Phase				Reproductive Phase								

2.1.3 Nutritional requirements (NPK)

Sunflower is a very demanding crop from the nutritional point of view, and its response to fertilization is limited by its productive potential and by the nutrient export rate. It is considered a crop that improves soil fertility because of its high recycling of nutrients to the detriment of the low export rate; it is a species sensitive to acidity, generally presenting symptoms of aluminum (Al) toxicity at pH (CaCl2) lower than 5.2 (CASTRO & OLIVEIRA, 2005).

The nutritional demand of sunflower varies according to the phenological stage, with the vegetative phase (up to 30 days after emergence - DAE), the one with the lowest demand and from this point until full flowering (R 5.5 phase) there is greater development of the crop (CASTRO & OLIVEIRA, 2005). These authors also add that from 28 to 56 days DAE there is a rapid increase in nutrient demand and in the phases of flowering and beginning of the filling of the achenes (R5, R6 and R7); between 56 and 84 days, there is a gradual reduction in the speed of nutrient uptake when the maximum level of accumulation is reached in variable quantities for each nutrient. Zobiole et al. (2010) mention that until 28 days after sowing the accumulation of nitrogen (N) and phosphorus (P) is small; from this point the rate of accumulation of N and P increases sharply reaching the point of greatest uptake at 85 days, for both nutrients. Potassium (K), in turn, reaches the absorption peak at 74 DAE - R7 stage, being the nutrient required in larger quantities by the sunflower. These authors also add that, because of the dynamics of N and K, it is recommended to perform the top dressing fertilization of these nutrients at approximately 30-35 DAE, so that there is adequate availability during the critical phase of accumulation.

According to Castro & Oliveira (2005) sunflower accumulates a total of 41 kg N, 17.1 kg P2O5 and 171 kg K2O to produce one ton of grains. Moreover, Prado & Leal (2006) complement that N, P and K are the elements that most limit the growth and dry matter production of this crop, however, about 90 to 95% of the amount of K absorbed can return to the soil with the mineralization of crop residues (ZOBIOLE et al., 2010). For Campos et al. (2012) fertilization with 60 kg ha-1 of N; 80 kg ha-1 of P2O5 and 80 kg ha-1 of K2O provides better quality of sunflower inflorescences cv. Embrapa 122/V2000.

N is one of the most important elements for sunflower cultivation; its deficiency significantly reduces the biometric characteristics of the plant such as height, number of leaves and stem diameter, also presenting symptoms characteristic of the deficiency of this nutrient (PRADO & LEAL, 2006). This element is essential for the synthesis of chlorophyll and is directly linked to photosynthesis, since it is part of the chlorophyll molecule, vitamins and energy systems in the plant. In this sense, N is directly related to protein formation (DECHEN & NACHTIGALL, 2007). Nasim et al. (2011) found that increasing the dose of N provides an increase in grain yield and productive components of sunflower.

P is an essential element for plant reproductive organs, such as inflorescence, grain formation and maturation (OSMAN & AWED, 2010). The uptake of P is most intense in chapter formation and flowering (ROSSI, 1998). The omission of this nutrient reduces the dry matter production of leaves, stem and roots of sunflower, also affecting the biometric characteristics of growth (PRADO & LEAL, 2006). According to Castro & Oliveira (2005) for a production of 2,000 kg ha-1 the minimum P export fertilization can vary from 30 kg ha-1 to 80 kg ha-1 of P_2O_5, respectively, in soils with high and very low phosphorus content. According to Ali & Noorka (2013) the combination135 kg ha-1 of N and 75 kg ha-1 of P promotes higher sunflower achene yield under irrigated cultivation in Sargodha-PK.

K is a very mobile element in the plant, with high mobility intracellularly and in tissues, traveling long distances with distribution by the vessels of the xylem and phloem (MEURER, 2006); it acts in the regulation of cellular osmotic potential and is linked to the balance of negative charges of intracellular organic acids and the balance of anions absorbed by the roots (BERNARDI et al. , 2008). It is also required for the activation of many enzymes of respiration and photosynthesis (TAIZ & ZEIGER, 2013). For Amanullah & Khan (2010), the combined application of 100 kg K and 45 kg P ha-1 provides increment in production components and grain yield of sunflower cv. Hysun-33 in northwest Pakistan. According to these authors, very low levels of K and P promote anticipation of the flowering phase, duration of the flowering phase and maturation of achenes, due to the lower rate of N uptake.

2.2 Biochar

2.2.1 General aspects

Biochar is a product resulting from the process of fast or slow pyrolysis of organic compounds, such as cattle manure, sewage sludge and crop residues, with the purpose of soil application (MAIA et al., 2011). Pyrolysis consists of the process of decomposition of plant material through the action of optimal temperatures between 450 to 550 °C in the absence or low concentrations of O_2 (LEHMANN, 2007), and the process of slow pyrolysis provides higher yield of biochar compared to fast pyrolysis; however, the amount of hours to perform the production process is higher. At these temperatures, biomass undergoes thermal decomposition, releasing large amounts of volatile components and heat, producing a carbonaceous matrix with a rough porous structure (CZERNIK & BRIDGWATER, 2004), and hydrocarbon-rich gas and bio-oil, which are used in the production of electricity, biofuels and hydrogen for domestic use, and in cars.

The physical-chemical properties of biochar were first scientifically observed in the Amazon basin, in soils known as "Terra Preta de Índio" (TPI) or anthropic dark earth. According to Falcão et al. (2001) these soils are characteristic of the Amazon region and formed by anthropic action, which present high levels of calcium oxide (CaO) and available P, high levels of organic matter, and intense biological activity from, probably, human and animal bone remains. Since the initial investigations of TPI in the 1990s and early 2000s, in recent years, research on biochar as a soil amendment for agricultural purposes has been intensified.

Biochar has high organic carbon content and large surface area due to its porosity as a result of the burning process of organic matter (MADARI et al., 2009). These characteristics make this technology with potential to be used in worldwide problems such as soil degradation, food and fertilizer shortages, competition for biomass and escalating greenhouse gas emissions (REZENDE et al., 2011). Woolf et al. (2010) estimated that annual net emissions of $_{CO2}$, N2O and CH4 could be reduced by 12% with biochar implementation without endangering food security, habitat or soil conservation. Wang et al. (2011) further point out that N2O and $_{CO2}$ emissions from rice-growing soils reduced substantially with the incorporation 50 t ha-1 of biochar, due to reduced NH4+ and NO3- concentrations in the soil.

Despite being produced from various sources of biomass, biochar can differ in their physicochemical properties due to the pyrolysis conditions employed in its production, such as: final pyrolysis temperature, carbonization rate and duration of carbonization (MUKHERJEE & LAL, 2013); however,

Lehman & Joseph (2009) point out that its high carbon content and its strong aromatic structure, are constant characteristics. Such characteristics are largely responsible for its chemical stability. Following this reasoning there is also little variability in pH between different types of biochar, being typically greater than seven. It is noteworthy that biochar has a wide variety of mineral nutrients, such as N, P, K, S, and can be used as a soil amendment, improving soil fertility and providing higher crop yields.

2.2.2 Biochar in soil

The use of biomass for energy production requires soil resources, in terms of land area and soil productivity. Biochar production technologies are unique among available biofuel technologies that can contribute to the improvement or maintenance of soil properties and thus to sustainable energy and food production (MADARI et al., 2012). Maia et al. (2011) comment that the contribution of biochar addition to soil encompasses two relevant aspects.

The first aspect concerns the beneficial effects on the chemical, physical and biological properties of the soil, besides providing greater biomass accumulation and crop productivity (GLASER et al, 2002;. KOOKANA et al, 2011). Madari et al. (2006) emphasize that the addition of eucalyptus charcoal positively affects soil properties by reducing potential acidity and increasing the P and K content. According to these authors, the use of charcoal in upland rice provided better initial vigor and biomass accumulation, more uniform development and greater seed dry mass. Furthermore, the use of biochar can reduce or replace the use of nitrogen fertilizers, responsible for increasing the concentration of nitrous oxide in the atmosphere. According to Lehmann et al. (2003) and Steiner et al. (2007) the application of biochar becomes more efficient when associated with mineral fertilization because of its low availability of nutrients; these authors also report that biochar retains nutrients, reducing their loss by leaching and increasing plant uptake.

The second aspect refers to its contribution to soil carbon sequestration because it has in its composition, stable forms of organic carbon, increasing its residual effect on the soil (SANTOS et al. , 2012). According to Fox & Chapman (2011), the production and dispersion of biochar in soils are considered a viable strategy to reduce global warming. Yanai et al. (2007) also point out that biochar and co-production of bioenergy through forest biomass can favor the change of the global climate context, reducing the use of fossil fuels by sequestering carbon in stable soils creating carbon reservoirs and drastically reducing emissions of nitrous oxide, a greenhouse gas more harmful to the environment than CO_2.

It is not known yet, the ideal amount of biochar for good growth and plant development; however, this amount should be determined depending on the type of soil, the plant and also the composition of biochar; however, according to Jien & Wang (2013) the application of 5% biochar improves soil physicochemical properties, notably of acidic soils and reduces erosion losses from highly weathered soils, the former due to significant increases in pH, cation exchange capacity, base saturation and microbial biomass and the latter mainly due to the formation of micro aggregates. Matovic (2011) took as a value for the study he developed the average of 3% of biochar in the first 30 cm of soil column, which leads to a dosage of 13.5 t ha-1; furthermore, Thomas et al. (2013) cite that the application of biochar to soil can ameliorate the effects of salinity on plants, through sorption of accumulated salt suggesting new applications to mitigate the effects of salinization in agricultural production, urban and contaminated soils.

2.2.3. poultry litter biochar

Among the different biomass sources used in the production of biochar, chicken litter stands out in Brazil. A mixture of substrates (known as "bed") of feces, feathers and leftover feed distributed in a barn or stable, to serve as bedding for animals, and commonly used industrial by-products or agricultural crop residues, such as: marvalha, waste from industrial processing of wood, shredded corn cob, rice husks, crop straw in general, grass hay and peanut shells (ÁVILA et al., 1992). Chicken litter is considered a good source of nutrients, especially N and P, and can reduce the use of mineral fertilizers. Moreover, due to the increase of organic matter to the soil, the use of poultry litter improves the physical attributes, favoring greater water retention, erosion reduction, improvement of aeration and development of soil microbial flora (BLUM et al., 2003); but, depending on the origin of the chicken litter (broilers or laying hens) and the number of layers of shavings, the contents of N, P, K, Ca and Mg may vary (GIANELLO & ERNANI, 1983; MIELE & MILAN, 1983; SCHERER, 1995). Costa et al. (2009) emphasize that chicken litter is an alternative of great viability in agricultural use because it is available on farms at low cost. Dikinya & Mufwanzala (2010) reinforce that soil fertility can be improved by the application of chicken manure because of the increase of exchangeable bases in the soil.

It is estimated that the annual Brazilian production of chicken litter is around 6.8 million ^{m3} (CORRÊA & MIELE, 2011). According to Santos et al. (2013) the use of poultry litter biochar promotes soil fertility improvement and can be used as a complementary source of phosphorus for crops. Sanvong & Suppadit (2013) cite that poultry litter biochar can be used as a fertilizer and an effective soil conditioner, being produced in a simple way and in a low-cost system, favoring self-sustainable agriculture. Corroborating, Utomo et al. (2012) verified improvements in soil fertility especially increased C.O, N, K and CTC content of the soil with the application of poultry litter biochar. In this sense, studies involving the use of poultry litter biochar in production systems stand out in importance, since the application ranges are variable, depending on the type of soil, the quality of the waste and the crop used.

MATERIAL AND METHODS

3.1 Location and characterization of the experimental area

The experiment was developed between the months of September and December 2014, in a protected environment belonging to the Academic Unit of Agricultural Engineering of the Center for Science and Technology and Natural Resources (CTRN), Federal University of Campina Grande (UFCG), located in the geographic coordinates of 7°13'11" south latitude and 35°53'31" west longitude and altitude of 547.56 m.

The region's climate, according to Köeppen's climate classification, adapted to Brazil (COELHO & SONCIN, 1982), is of type Csa, which represents a mesothermal, sub-humid climate, with a hot and dry dry dry season (4 to 5 months) and a rainy season from autumn to winter.

The climatic data regarding maximum, minimum, and relative humidity in the environment outside the greenhouse were obtained from the weather station of Campina Grande (Figure 1), with average maximum and average minimum temperatures of 27.40 and 19.00°C, respectively, and relative humidity ranging from 77 to 87%.

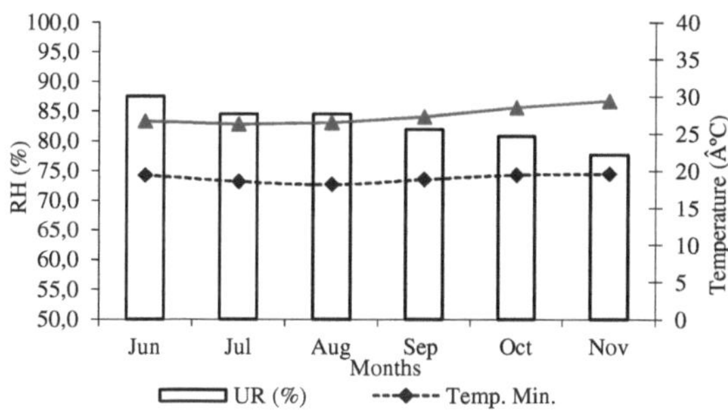

Source: AESA (2014)

Figure 1 - Maximum and minimum temperature and relative humidity during the experiment

3.2 Experimental design

The experimental design was entirely randomized, with treatments arranged in a 5 x 4 factorial scheme, with four repetitions, totaling 20 treatments and 80 experimental units, being the treatments composed by the combination of five doses of mineral fertilization [D1=0, D2=25%, D3=50% D4=75% and D5=100%] of the indication of fertilization with NPK (100; 300 and 150 mg kg-1) for trials according to Novais et al. (1991) and four doses of poultry litter biochar [0; 5% (400 g/vase); 10% (800 g/vase) and 15% (1100 g/vase)], calculated based on soil volume.

3.3 Cultivar

Sunflower seeds, cultivar Embrapa 122/V2000, provided by Embrapa Algodão in Campina Grande were used. According to Embrapa (2006) this genetic material stands out for its precocity, productivity, quality and oil content in the acheneia, which has an average vegetative cycle of 100 days, with the beginning of flowering at 53 days, physiological maturation at 85 days, plant height of 155 cm, oil content ranging from 40-44%, mass of one thousand acheneia of 60 g and average productivity of 1741 kg ha-1.

3.4 Installation and conduction of the experiment

For sowing and planting, plastic containers of 20 dm3 capacity were used (Figure 1A), filled with 20 kg of soil, leaving 5 cm free on the surface of the pot to facilitate irrigation. The pots were perforated at the base with five holes of 8 mm in diameter to facilitate drainage. We used non-saline and non-sodic sandy soil collected at a depth of 0-30 cm properly sifted and from the municipality of

Campina Grande, PB, whose physical and chemical attributes were determined according to the methodology proposed by Embrapa (2011) (Table 2).

The pots were distributed on platforms made with ceramic blocks, at a height of 0.2 m, distributed in single rows spaced 0.5 m apart and 0.4 m between plants within the row (Figure 1B). After packed in the pots, the soil material was brought to field capacity by the capillary saturation method, followed by free drainage.

Figure 2 - Pot used in the experiment (A) and overview of the experimental area (B)

Table 2 - Physical and chemical attributes of the soil used in the experiment

Attributes	**Unit**	**Value**
Textural classification	-	Free Sand
Soil Density	g cm-3	2,71
Porosity	%	52,03
Assortative complex	**Unit**	
Calcium (Ca2+)	cmolckg-1	2,10
Magnesium (Mg2+)	cmolckg-1	2,57
Sodium (Na+)	cmolckg-1	0,06
Potassium (K+)	cmolckg-1	0,14
Hydrogen (H+)	cmolckg-1	1,78
Aluminum (Al3+)	cmolckg-1	0,00

CTC	cmolckg-1	6,65
Organic Carbon	g kg-1	5,5
Organic Matter	mg kg-1	9,5
Assimilable Phosphorus (P)	mg kg-1	45,0
Nitrogen	mg kg-1	0,5
pH H2O	-	6,00
CE	dS m-1	0,16
PST	-	0,90
Saturation Extract	**Unit**	
pHps	-	5,80
CEes	dS m-1	0,22
Chloride (Cl-)	mmolcL-1	1,00
Carbonate (CO32-)	mmolcL-1	0,00
Bicarbonate (HCO3-)	mmolcL-1	1,20
Calcium (Ca2+)	mmolcL-1	0,25
Magnesium (Mg2+)	mmolcL-1	1,50
Sodium (Na+)	mmolcL-1	0,70
Potassium (K+)	mmolcL-1	0,19
RAS	-	0,75

pHPS - Saturation paste pH; ECes- Electrical conductivity of the saturation extract; TSP- Percentage of exchangeable sodium; RAS- Sodium adsorption ratio

For fertilization, urea, Monoammonium-phosphate (MAP) and potassium chloride (KCl) were used as sources of NPK, respectively, applied in coverage, being applied per pot in the treatment with 100% of the recommendation 1.41 g of urea, 13.64 g of MAP and 5 g of KCl (Table 3). MAP was applied every 7 days starting at 25 days after sowing (DAS). KCl was applied at 30, 45 and 60 DAS. Urea was applied at 50 DAS. The biochar was incorporated into the soil and left to incubate for 20 days. Foliar fertilization was applied at 40 DAS using Ubyfol® at a proportion of 0.5 kg of foliar fertilizer to 100 L of water, distributed on the plants using a knapsack sprayer.

Table 3 - Urea, MAP and KCl values applied in each treatment

Dose (% of recommendation)	Urea	MAP	KCl
	-------------------	g/vaso	--------------------
0	0	0	0
25	0,35	3,41	1,25
50	0,71	6,82	2,50
75	1,06	10,23	3,75
100*	1,41	13,64	5,00

Fertilizer recommendations N: 100 mg kg-1; P: 300 mg kg-1; K: 150 mg kg-1 (NOVAIS et al., 1991)

The biochar was produced from the conventional pyrolysis process using, as residue, chicken litter (450ºC x 0.5 hour x atmospheric pressure), which was supplied by the company SPPT Pesquisas Tecnológicas LTDA, whose chemical attributes were determined according to the methodology proposed by Embrapa (2009) (Table 4).

Table 4 - Chemical attributes of the biochar used in the experiment

Feature	Value
pH H2O (1:2.5)	10,1
Macronutrients (g kg-1)	
Nitrogen (N)	42,31
Phosphorus (P)	32,56
Potassium (K+)	48,56
Calcium (Ca2+)	57,75
Magnesium (Mg2+)	12,40
Sodium (Na+)	14,37
Micronutrients (mg kg-1)	
Iron (Fe)	137
Copper (Cu)	812
Zinc (Zn)	700
Manganese (Mn)	862

Sowing was performed on October 19, 2014, placing three seeds per pot at 0.05 m depth and distributed equidistantly. Seedling emergence began on the fifth day after sowing (DAS) and continued until the fifteenth day. Thinning was done at 10 and 20 DAS leaving, respectively, two and one plant per pot, i.e. those with the best vigor.

Pollination was performed when the plants reached the R5 stage using artificial cross-pollination, which consisted in using a brush with a good amount of pollen from several flowers produced by different plants, with a quick touch in order to reach all the stigmas (Figure 3).

Figure 3 - Artificial cross-pollination

During the course of the experiment, the following cultivation practices were performed: manual weed elimination; staking of the plants when they reached the R6 stage, which is the end of flowering, and surface scarification of the soil when it presented some degree of surface crusting.

At 30 and 50 DAS the whitefly (*Bemisia tabaci*) was chemically controlled with endosulfan insecticide at a dose of 500 g i.a. ha-1.

3.5 Irrigation Management

Irrigation was measured by the water consumption of the plants under 100% ETr. The volume applied (Va) per pot was obtained by the difference between the average weight of the containers in the condition of maximum water retention (Pcc), which was determined by saturating the containers with water and draining them; when the drained volume was reduced, the containers were weighed, obtaining the Pcc value and the average weight of the containers in the non-saturation condition (current weight) (Pa), divided by the number of containers (n), as indicated in Eq. 1. The irrigations were performed daily, in the late afternoon, using rainwater, which was collected by the collection system of the greenhouse and stored in a box with a capacity of 5 ^{m3}.

$$Va = \frac{Pcc - Pa}{n} \quad (1)$$

3.6 Harvesting and processing

The harvest was performed when the plants reached the R9 stage, that is, when they presented achenes with hard mass, the chapter is inclined downwards, with yellow to brown back and bracts (SCHNEITER & MILLER, 1981). At this time the plants were cut close to the ground, separated into stem, leaf and chapter and placed in kraft paper bags, which were then placed in a forced air circulation oven at a temperature of 65°C until they reached constant mass. The achenes from each chapter were threshed manually and then separated into viable and non-viable achenes.

3.7 Variables analyzed in the soil

After the end of the experiment, Ca2+, Mg2+, Na+, K+, CO3-, HCO3-, SO4-, Cl-, EC and pH levels of the saturation extract were analyzed, according to the methodology proposed by Embrapa (2011).

3.8 Variables analyzed in sunflower

3.8.1 Growth variables

At 30, 60 and 90 DAS, plant height (PA), stem diameter (CD) and number of leaves (NF) were evaluated. The leaf area (LA) was evaluated at 30 and 60 DAS. The PA was measured in cm, from the neck of the plant to the insertion of the youngest leaf. The CD was measured at the neck of the plant at 2 cm from the ground (mm). In counting the number of leaves we considered those that were fully expanded with a minimum length of 3 cm; The AF in cm2, measured by the non-destructive method, using the methodology proposed by Maldaner et al. (2009), according to eq. 2.

$$AF = 1.7582 \text{ x } L^{1,7067} (2)$$

In which,

LA= leaf area (cm2);

L= leaf width (cm);

3.8.2 Absolute and Relative Growth Rates

With the data obtained from Plant Height and Stem Diameter in each season, the Absolute Growth Rate and the Relative Growth Rate were determined using the methodology proposed by Fernandes (2002) and Benincasa (2003), according to the equations described below:

For the absolute growth rate

$$TCA = \frac{(A_2 - A_1)}{t_2 - t_1}$$ (cm day-1or mm day-1)

For the relative growth rate

$$TCR = \frac{(\ln A_2 - \ln A_1)}{t_2 - t_1}$$ (cm cm-1 day-1 or mm mm-1 day-1)

where:

A2 = Height or diameter obtained at the end of the study period;

A1 = Height or diameter obtained at the beginning of the study period;

t2 - t1 = time difference between samplings.

3.9 Phytomass production

The variables stem dry phytomass - SCF, leaf dry phytomass - FSF, aboveground dry phytomass - FSPA, and chapter dry phytomass - FSCap, were evaluated after the harvest.

3.10. Production variables

The production components were evaluated at harvest, as follows: Chapter diameter (DCap), measured in cm at harvest; number of achenes per chapter (NAC), Number of viable achenes (NAV); percentage of non-viable achenes (PANV) - considering as non-viable achenes, the shocked grains; mass of one hundred achenes (M100A) - considering only the viable achenes and determined according to the methodology proposed by Brazil (2009) and production of achenes per plant (PROD).

3.11. Analysis of the results

The data obtained were evaluated by variance analysis using the 'F' test at 0.05 and 0.01 probability levels and, in cases of significance, linear and quadratic polynomial regression analysis was performed using the statistical software SISVAR (FERREIRA, 2011). Due to the heterogeneity of the data, it was necessary to transform into $\sqrt{X}$ or $\sqrt{X} + 1$.

RESULTS AND DISCUSSION

4.1. Soil analysis

The application of poultry litter biochar in the soil of the experimental units, promoted an increase of pHps and saturation extract electrical conductivity (ECes), and the highest values were observed with doses of 800 and 1100 g/vac of biochar (Table 5) which, according to Richards (1954) are characteristic values of saline soils (pH ≤8.5; ECes>4.0 dS m-1)· This behavior was corroborated by Liesch et al. (2010) who observed an increase in soil pH with increasing doses of poultry litter biochar after 28 days of incubation, obtaining maximum values of 8.87 with the application of 90 t ha-1. Similarly, Revell (2011) verified a linear increase in the accumulation of soluble salts in the soil with increasing doses of poultry litter biochar. It is also observed that the levels of chloride (Cl-), bicarbonate (HCO3-), sodium (Na+) and potassium (K+) followed the same trend of ECes, which had an increase of respectively, 33.94%, 65.58%, 51.08% and 91.81% in relation to the soil that did not receive fertilization with biochar. Such results, that is, high concentration of soluble salts in the experimental units, show the probable salinization of the soil due to fertilization with biochar (HOLANDA et al., 2010); however, there was a reduction in calcium (Ca2+) and magnesium (Mg2+) content in the saturation extract with increasing doses of biochar, a fact that influenced the increase of the sodium adsorption ratio (SAN) although the values in the experimental units (Table 5) were classified as low (< 13).

It is likely that the reaction of biochar, rich in calcium, magnesium, sodium and potassium (Table 4), in the soil, may have caused cation exchange, i.e., increased the adsorption of calcium and magnesium releasing sodium and potassium from the sorptive complex into the soil solution; another reason for the decrease of calcium and magnesium in the soil solution is the fact of the uptake of these elements by the crop.

According to Pessoa et al. (2010), the increase in the ASR can promote nutritional imbalance, due to the difficulty of Ca^{2+}, Mg^{2+} and K^+ uptake because of the increase in the proportion of soluble sodium in the soil solution. Furthermore, if the proportion of Na^+ in the soil solution continues to increase, despite its lower selectivity, it can replace the Ca^{2+} and Mg^{2+} of the sorptive complex by passing soluble sodium to exchangeable sodium, which can initiate the sodification process (RIBEIRO et al., 2003).

Table 5 - Average values of chemical attributes of soil cultivated with sunflower as a function of doses of poultry litter biochar

Attributes Saturation extract (mmolc L-1)	B1	B2	B3	B4
pHps	5,98	7,07	7,90	8,28
ECes - dS m-1	3,31	7,28	9,50	9,32
Chloride (Cl^-)	33,32	47,73	50,44	50,43
Carbonate (CO_3^{2-})	0,00	0,00	0,00	0,00
Bicarbonate (HCO_3^-)	2,44	2,95	5,36	7,09
Sulfate (SO_4^{2-})	P	P	P	P
Calcium (Ca^{2+})	7,04	7,63	3,82	3,97
Magnesium (Mg^{2+})	28,13	35,44	26,98	22,84
Sodium (Na^+)	16,98	31,03	37,01	34,71
Potassium (K^+)	4,23	24,11	44,00	51,66
RAS	4,05	6,69	9,43	9,48

B1= 0; B2 400 g/vessel; B3 = 800 g/vessel; B4 = 1100g/vessel; pHPS - pH of the saturation paste; ECes- Electrical conductivity of the saturation extract; RAS- Sodium adsorption ratio

4.2 Growth variables

4.2.1 Plant height and stem diameter

The application of increasing doses of NPK in the experimental units exerted a significant effect on plant height (PA) at 90 days after sowing sunflower (DAS) and stem diameter at 30, 60 and 90 DAS (Table 6). Such results may be related to the low nutritional demand of sunflower at the beginning of the crop cycle, especially in the first 30 DAS. According to Zobiole et al. (2010) the greatest uptake of nitrogen (N) and phosphorus (P) by sunflower occurs from 28 days after emergence (DAE), reaching the point of greatest uptake at 85 DAE; potassium (K), in turn, reaches the peak of uptake at 74 DAE

Table 6 - Summary of the 'F' test for plant height (PA) and stem diameter (CD) of sunflower at 30, 60 and 90 days after sowing (DAS) as a function of doses of NPK and poultry litter biochar

Source of Variation	F Test					
	AP (cm)			DC (mm)		
	30 DAS	60 DAS	90 DAS	30 DAS	60 DAS	90 DAS
Doses (D)	ns	ns	**	*	**	**
Linear Reg.	-	-	**	ns	ns	ns
Quadratic Reg.	-	-	*	*	**	**
Biochar (B)	**	**	ns	**	**	ns
Linear Reg.	**	**	-	**	**	-
Quadratic Reg.	ns	ns	-	*	ns	-
D x B	ns	ns	ns	ns	**	*
CV (%)	33,72	20,79	16,80	27,05	17,54	18,90

(**), (*); (ns) significant at ($p\leq 0.01$) and ($p\leq 0.05$) and not significant respectively

Biochar significantly affected the AP and CD at 30 and 60 DAS. Significant effect of the interaction between D x B factors is found for DC at 60 and 90 DAS (Table 6).

According to the regression equation (Figure 4A) there was a quadratic effect on the AP of sunflower at 90 DAS as a function of the NPK dose, and an increase in AP (97.45 cm) up to a dose of 33.33% of the recommendation, which corresponds to an increase of 3.88% in relation to the lowest dose. It is also observed that from this point on there was a reduction of 15.48%; such results show that under the development conditions of the research the fertilization recommendation adopted for sunflower was much higher than the needs of the crop. Chaves et al. (2014) found no significant effect of N and K fertilization on sunflower PA; however, P fertilization promoted a linear increase in PA obtaining a maximum value at 60 DAS of 110.47 cm, with the dose 120 kg ha-1of P2O5. According to Jahangir et al. (2006) the combination of 120 kg ha-1 of N and 75 kg ha-1 of P2O5 promoted greater height of sunflower plants, obtaining a maximum height of 62.1 cm. Salih (2013) also verified, evaluating the combination of different doses of N and P in Sudan, increments in sunflower plant height with increasing doses of mineral fertilization. Uchôa et al. (2011) found significant effect of potassium fertilization for AP of sunflower cv. Embrapa 122/V2000 and obtained a maximum height of 126 cm with the application of 78.6 kg ha-1 of K2O.

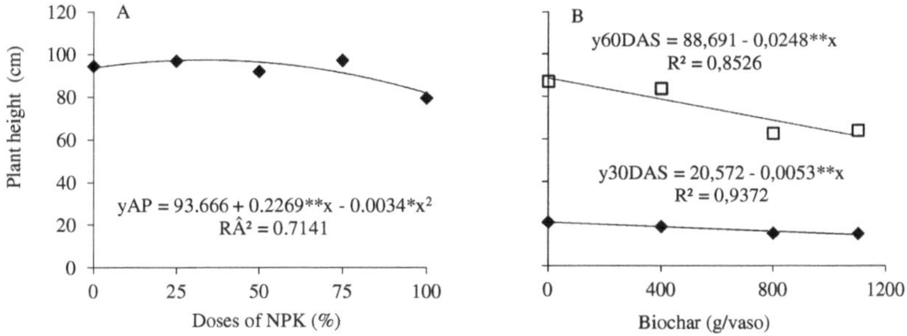

Figure 4 - Sunflower plant height at 90 DAS as a function of NPK (A) and chicken litter biochar (g/vaseo) doses at 30 and 60 DAS (B)

The results for the AP as a function of the doses of biochar were best fitted to the linear model (Figure 4B) indicating that the increase in the doses of biochar promoted decreases in the AP at 30 and 60 DAS. The lowest estimated values of AP for plants grown with fertilization of 1100 g/vessel, at 30 and 60 DAS were, respectively, 14.74 and 61.41 cm, which corresponded to a reduction of 28.33% and 30.76% compared to plants that were not fertilized with biochar. Such results may be related to the increase in soil salinity evidenced by the high soil EC values as mentioned in Table 5; moreover, excess salts in the soil can reduce N uptake by plants, while the levels of Cl-absorbed and accumulated are increased, promoting less growth (BOSCO et al., 2009). Nobre et al. (2010) and Moraes et al. (2011), found linear reduction of sunflower PA under salt stress conditions. According to Suppadit et al. (2012), the application of biochar at levels above 98.4 g/10 kg of soil limits soybean growth due to its alkaline nature, promoting pH increment to levels above that recommended for the crop. However, Utomo et al. (2012) found an increase in the height of corn plants with the combination of 145 kg ha-1 of N and 50 t ha-1 of poultry litter biochar compared to the control (no fertilization).

Figure 5 shows the results of sunflower CD at 30 DAS as a function of the doses of NPK and chicken manure biochar.

According to the regression equation (Figure 5A) there was a quadratic effect on the CD of sunflower at 30 DAS as a function of the NPK dose. The dose of 42.83% of the recommendation (4.25 mm) provided an increase of 13.64% in relation to the lowest dose (3.67 mm). It is also observed that from this point (4.25 mm) there was a reduction of 23.76% in relation to the highest dose, which provided the lowest CD values (3.24 mm). Chaves et al. (2014) found significant effects of N and K fertilization on sunflower stem diameter; however, phosphorus had no significant effect on this variable. Siddiqui et al. (2009) found an increase in sunflower stem diameter with increasing mineral fertilization, with the highest values being recorded with a combination of 90:45:45 kg ha-1 of NPK.

The increase in biochar dose promoted a linear reduction in the CD of sunflower at 30 DAS (Figure 5B). The lowest CD value was 2.99 mm when a dose of 1100 g/vessel was applied, which corresponded to a decrease of 39.84% in relation to plants that did not receive fertilization with biochar (4.97 mm). This ratifies the hypothesis cited above, that the increase in soil salinity due to fertilization with biochar also restricted the growth in stem diameter of sunflower. According to Ayers & Westcot (1999), the presence of specific ions in the soil solution causes an increase in its osmotic pressure, negatively affecting physiological processes, reducing water uptake by the roots, inhibiting meristematic activity and cell elongation, resulting in a reduction in plant growth. Nobre et al. (2010) verified a linear reduction of CD of sunflower cv. Embrapa 122 V/2000 under salt stress conditions, obtaining an average diameter of 6.05 mm at 26 DAS.

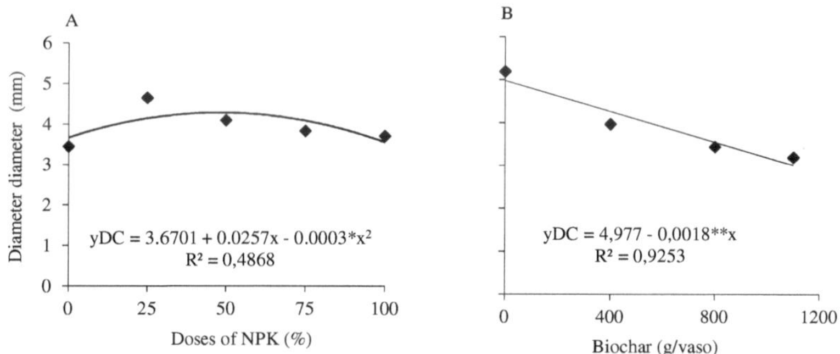

Figure 5 - Sunflower stem diameter at 30 DAS as a function of NPK doses, % (A) and chicken litter biochar, g/vessel (B)

The interaction between factors (D x B) for CD at 60 and 90 DAS can be seen in Figure 6. At 60 DAS (Figure 6A) and according to the regression equations (Table 7) there was a quadratic effect in plants subjected to fertilization with 0 and 400 g/vasion of biochar, with an increase in CD up to a dose of 55.68% (11.68

mm) and 46.83% (10.05 mm) of the recommendation, respectively for doses of 0 and 400 g/vasion.

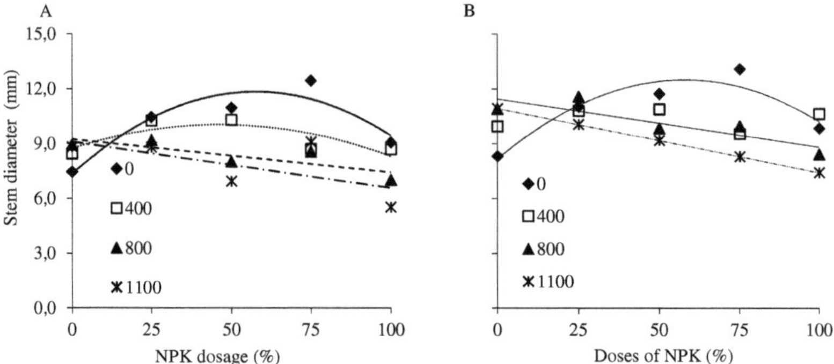

Figure 6 - Stem diameter at 60 (A) and 90 DAS (B) as a function of NPK - % and chicken litter biochar - g/vaseo

However, a linear decreasing effect was observed in plants fertilized with 800 and 1100 g/vessel of biochar (Figure 6A). According to the regression equations (Table 6), there was a reduction of 19.38% and 27.52%, respectively, when comparing the lowest and highest doses of NPK fertilization.

Table 7 - Regression equations for sunflower stem diameter at 60 and 90 days after sowing (DAS) as a function of NPK and chicken manure biochar doses

	Stem diameter at 60 DAS (mm)		
Biochar (g/vessel)	Equation	R2	Average
0	Y = 7.3487+ 0.1559*x - 0.0014**x2	0,8763	10,08
400	Y = 8.7411+ 0.0562x - 0.0006*x2	0,6252	9,28
800	Y = 9.2338 - 0.0179*x	0,6711	8,34
1100	Y = 9.085 - 0.025*x	0,4099	7,83
	Stem diameter at 90 DAS (mm)		
Biochar (g/vessel)	Equation	R2	Average
0	Y = 8.1634+ 0.1497x - 0.0013**x2	0,8830	10,81
800	Y = 11.461 - 0.0263**x	0,7565	10,14
1100	Y = 10.939 - 0.0352**x	1,0000	9,18

(**), (*) significant at ($p \leq 0.01$) and ($p \leq 0.05$); Y = stem diameter (mm); x = NPK doses (%)

Regarding the CD, at 90 DAS (Table 6) there was a better fit of the data to the quadratic model when the plants did not receive fertilization with biochar (Dose 0) (Figure 6B), whose dose of 57.58% (12.47 mm) of the recommendation of NPK promoted an increase of 34.54% in reference to the lowest dose and from this point there was a reduction of 18.74% in relation to the dose of 100% of the recommendation of NPK; however, these values are lower than those observed for this cultivar (18.4 mm) (SMIDERLE et al, 2005; BÍSCARO et al., 2008). The doses of 800 and 1100 g/vessel of biochar promoted linear reduction of CD at 90 DAS (Figure 6B) and according to regression equations (Table 7) there was a reduction of, respectively, 22.96% and 32.19% when compared to the lowest and highest dose of NPK recommendation. In relation to the plants that received the dosage of 400 g/vessel of biochar, according to regression studies (Figure 6B) there was no significant effect on CD, reaching an average value of 10.37 mmbefore which it is observed that fertilization with biochar presents the same trend observed for CD at 30 DAS (Figure 5), a fact linked to the effects of salt stress, as mentioned earlier (Table 5); however Carvalho et al. (2011) found positive effects of fertilization with chicken litter on the growth and yield of soybeans, which leads to believe, according to Mukherjee & Lal (2013) that the nature of the material used in the production of biochar and the pyrolysis conditions used, favored the release of ions that promoted soil salinization, but Thomas et al. (2013) verified in a study with biochar produced from beechwood sawdust (*Fagus grandifolia*) that the application of 50 t ha-1 reduces the effects of salt stress on bean plants.

4.2.2 Leaf number and leaf area

The doses of NPK had a significant effect on the number of leaves per plant (LE) and leaf area (AF) at 60 DAS (Table 8). There was also a significant effect of the amount of biochar for the number of leaves and leaf area in all evaluation periods; however, there was no significant effect for the interaction D x B for any of the variables analyzed. These results are related to the low nutritional demand of sunflower in the vegetative phase. According to Castro & Oliveira (2005), the lowest nutritional demand of sunflower occurs until 30 days after emergence (DAE), with a gradual increase until 56 DAE.

Table 8 - Summary of the 'F' test for number of leaves per plant (LE) and leaf area of sunflower at 30 and 60 days after sowing (DAS) as a function of doses of NPK and poultry litter biochar

Source of Variation	F Test			
	NF		AF (cm2)	
	30 DAS	60 DAS	30 DAS1	60 DAS
Doses (D)	ns	*	ns	**
Linear Reg.	-	*	-	ns
Quadratic Reg.	-	*	-	**
Biochar (B)	**	**	**	**
Linear Reg.	**	**	**	**
Quadratic Reg.	ns	*	*	ns
D x B	ns	ns	ns	ns
CV (%)	18,65	19,08	29,59	31,99

(**), (*); (ns) significant at ($p \leq 0.01$) and ($p \leq 0.05$) and not significant, respectively
1statistical analysis performed after data transformation to √X

According to the regression equations, there was a quadratic effect for sunflower growth at 60 DAS as a function of the NPK doses (Figure 7A). It is also observed that up to a dose of 34.25% (18 leaves) of the fertilizer recommendation there was an increase in the NC of 3.89% in relation to the lowest dose (Dose 0) and from this point on there was a reduction of 14.32% in relation to the dose of 100% (N:P:K, 100:300:150 mg kg-1) (14 leaves) of the recommendation. The nutritional state of the plant, especially the adequate supply of N, P and K, is directly related to its growth and development because these elements have functions linked to energy metabolism. Fagundes et al. (2007) observed, evaluating the growth of sunflower as a function of doses and sources of N, a quadratic effect for this variable and obtained the maximum number of 24 leaves at the end of the cycle, whose results are in accordance with Chaves et al. (2014), who verified positive effects of NPK fertilization for this variable, obtaining maximum number of 16 leaves at 60 DAS with the combination of 71.22 kg ha-1 of N, 120 kg ha-1of P2O5 and 78.6 kg ha-1of K2O. Adebayo et al. (2010) found an increase in the number of leaves as a function of P doses, obtaining maximum number of leaves (35 leaves) with the application of 30 kg ha-1 of P (4.8 g per plant). Biscaro et al. (2008) found a maximum number of leaves per plant (29 leaves) with a dose of 80 kg ha-1 when sunflower growth was evaluated under nitrogen fertilization.

Fertilization with poultry manure promoted a linear reduction in the number of sunflower leaves at 30 and 60 DAS by 34.52% and 20.56%, respectively, when compared to plants that were grown without biochar fertilization (Dose 0) and the dose of 1100 g/vaseo (Figure 7B). The greatest number of leaves was obtained when the plants were grown without biochar fertilization (19 leaves), which was lower than the average number of leaves observed for this cultivar (29 leaves) (BISCARO et al., 2008). Once again, the negative effects of salt stress provided by fertilization with biochar from chicken litter are verified. According to Oliveira et al. (2013) the reduction of the NF functions as an anatomical adaptation in salt stress conditions, reflecting the reduction of transpiration as an alternative to maintain the absorption of water.

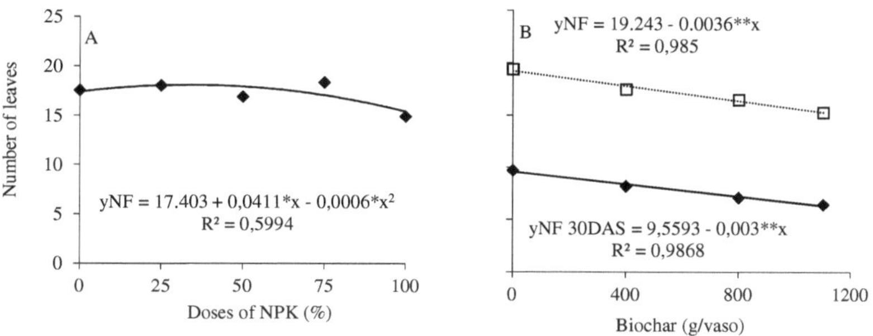

Figure 7 - Number of leaves at 60 DAS as a function of NPK doses - % (A) and at 30 and 60 DAS as a function of chicken litter biochar doses - g/vessel (B)

The dose of 49.15% (1285.83 cm2) of the recommended fertilization with NPK promoted, at 60 DAS, an increase in the leaf area (AF) of 28.90% in relation to the smallest dose (Dose 0) (914.22 cm2) and from this point on there was a reduction of 30.14% in relation to the plants that received the dose of 100% of the recommendation (898.22 cm2) (Figure 8A). The greater availability of nutrients provides greater plant growth and, consequently, greater expansion of the leaf lamina, but when the dose of 100% of the recommendation was applied there was a very sharp reduction in relation to the maximum estimated dose, a fact that may be related to the event of the fertilizer recommendation adopted, that is, based on the mass of soil in the pot, which may have provided a dose above the requirements of the crop. According to Ali & Noorka (2013) the combination135 kg of N and 75 kg ha-1 of P promoted increment of AF of sunflower in irrigated cultivation in Sargodha-PK, obtaining maximum AF of 5813.24 cm2.

A linear reduction of sunflower AF was observed at 30 and 60 DAS with increasing doses of biochar (Figure 8B), with the lowest values of respectively 60.59 and 717.95 cm2 with the dose and 1100 g/vaseo. It is also observed that the fertilization with biochar promoted a reduction of 76.90% and 52.29%, respectively, at 30 and 60 DAS, when comparing the lowest and highest dose. These results can be attributed to the effect of excess salts in the soil (Table 5), which, according to Taiz & Zeiger (2013) can promote changes in plant physiology, which are often accompanied by morphological and anatomical changes, such as a reduction in growth and leaf area, associated mainly with a reduction in turgor pressure, which interferes with the processes of elongation and cell division. According to Tester & Davenport (2003) the reduction of the leaf area is one of the adaptation mechanisms of the plant to salt stress, reducing the transpiring surface, keeping the water potential in the plant high. Nobre et al. (2011) found a linear reduction in the leaf area of sunflower with increasing salinity of irrigation water; however, Coomer et al. (2012) found that the combination of 3000 kg ha-1 of poultry litter biochar plus 56 kg ha-1 of N promoted an increase in the leaf area of cotton compared to the control treatment (no application of biochar + 56 kg ha-1 of N).

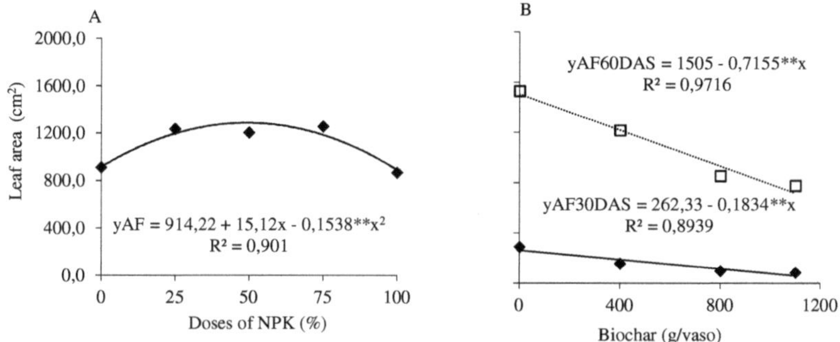

Figure 8 - Sunflower leaf area at 60 DAS as a function of NPK doses - % (A) and at 30 and 60 DAS as a function of chicken litter biochar doses - g/vessel (B)

4.2.3 Growth rates

The application of poultry manure in the experimental units had a significant effect on the absolute and relative growth rates of sunflower plant height in the periods evaluated from 30 to 60 DAS (TCAAP1 and TCRAP1) and 60 to 90 DAS (TCAAP2 and TCRAP2) (Table 9); however, there was no significant effect for the factor doses of NPK fertilization and interaction (D x B) for any variable analyzed.

Although the increase in mineral fertilization has promoted greater growth of sunflower plants, a fact observed by the higher values of AP and CD (Figures 4, 5 and 6), the increase in biomass as a function of time (growth rate) remained constant at all levels of fertilization. Ribeiro et al. (2014) found no significant effect of nitrogen fertilization on sunflower growth rates. Corroborating, Furtado et al. (2014) found no significant effect of potassium fertilization on TCAAP and TCRAP in the evaluation period from 35 to 60 DAS.

Table 9 - Summary of the 'F' test for absolute and relative growth rate of sunflower plant height in the evaluation periods carried out between 30-60 (TCAAP1 and TCRAP1) and 60-90 (TCAAP2 and TCRAP2) days after sowing (DAS) as a function of NPK and chicken manure biochar doses

Source of Variation	F Test			
	TCAAP1	TCAAP21	TCRAP1	TCRAP21
Doses (D)	ns	ns	ns	ns
Biochar (B)	**	**	*	**
Linear Reg.	**	**	*	**
Quadratic Reg.	ns	*	*	*
D x B	ns	ns	ns	ns
CV (%)	22,25	42,47	14,67	45,84

(**), (*); (ns) significant at ($p \leq 0.01$) and ($p \leq 0.05$) and not significant respectively
1statistical analysis performed after data transformation to $\sqrt{X}$

The increase in the amount of biochar promoted a linear reduction in the absolute growth rate of sunflower plant height in the period from 30 to 60 DAS (TCAAP1) (Figure 9A). The lowest TCAAP1 values were 1.48 cm day-1 when plants received fertilization with 1100 g/vessel of biochar which corresponded to a decrease of 34.25% compared to plants that were not fertilized with biochar (2.25 cm day-1). As for AP (Figure 4B), the effect of salt stress due to biochar fertilization also influenced TCAAP1. Ribeiro et al. (2014) found a reduction on sunflower growth rate in the period 31-45 DAS under salt stress conditions. Maia Junior et al. (2013) observed, on average, TCAAP of 1.31 cm day-1 in the evaluation period of 20-80 DAS for cv. Embrapa 122/V2000. The inhibition of plant growth under salt stress conditions occurs due to the reduction of the osmotic potential of the soil solution to levels that make the absorption of water and nutrients impossible, combined with ionic toxicity, nutritional imbalance or both, due to the excess accumulation of certain ions in plant tissues (FLOWERS, 2004).

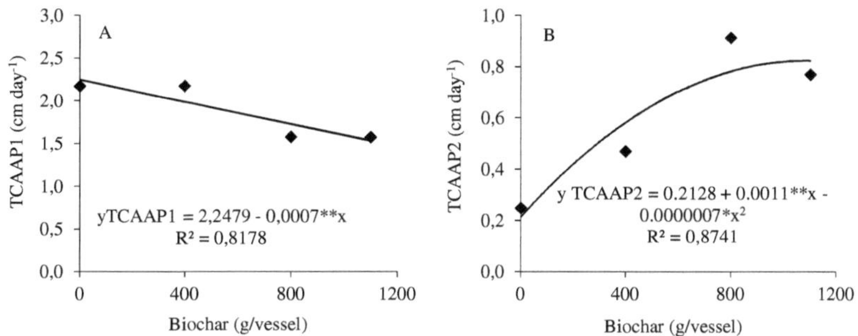

Figure 9 - Absolute growth rate of plant height at 30-60 DAS - TCAAP1 (A) and 60-90 DAS -TCAAP2 (B) as a function of doses of chicken manure biochar - g/vessel

The vegetative growth rates are variables used to determine plant growth at different stages of development, being dependent on the amount of material previously accumulated. Such variables represent increase in height or mass, as a function of time (absolute growth) or growth per unit of time, expressed on a common basis initial mass, initial area, initial length (relative growth) (TAIZ & ZEIGER, 2013). Therefore, through these variables, it is possible to determine the intensity of biotic or abiotic stress on plant growth.

However, based on the regression studies for absolute growth rate of sunflower plant height in the period from 60 to 90 DAS (TCAAP2) (Figure 9B) the best model was a quadratic model, indicating that the increase in the dose of biochar up to 785.7 g/vessel promoted an increase in TCAAP2 of 67.00% in relation to plants that were not fertilized and from this point there is a decrease of 10.71% in relation to the highest dose; These results can be justified by the greater sensitivity of the crop to the effects of salt stress provided by biochar in the vegetative phase, a fact that may be related to the greater adaptation of the crop after 60 DAS. Corroborating this, Ribeiro et al. (2014) found no significant effect of salt stress on the growth rates of sunflower in the evaluation periods 46-60 and 61-75 DAS; these authors also add that these results are related to permanent osmotic adjustment, which is a mechanism that allows halophytic plants to live under salt stress conditions.

According to the regression equations it was observed a quadratic effect for the relative growth rate of sunflower plant height in the period from 30 to 60 DAS - TCRAP1 (Figure 10A), in which the dose of biochar from 400 g/vessel promoted decrease of TCAAP1 of 9.72%; However, the relative growth rate of sunflower plant height in the period from 60 to 90 DAS - TCRAP2, increased linearly with the increase of doses of poultry manure biochar being the highest values obtained with the dose of 1100 g/vessel (0.0121cm cm-1 day-1), which promoted an increase of 72.73% compared to plants that did not receive fertilization with biochar (Figure 10B).

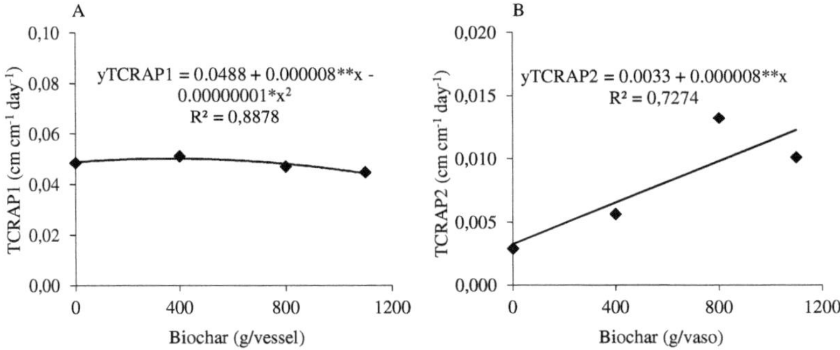

Figure 10 - Relative growth rate of plant height at 30-60 DAS - TCRAP1 (A) and 60-90 DAS -TCRAP2 (B) as a function of doses of chicken manure biochar - g/vessel

Based on these results, the relative growth rates followed the same trend as the absolute growth rates (Figure 9), indicating that after 60 DAS there was possibly a greater adaptation of the crop to stress conditions. Nobre et al. (2014) found a linear increase in the RGRAP of castor oil plant in the period from 14 to 120 DAS with increasing water conductivity. These authors mention that these results demonstrate the efficiency of the culture in producing new material per unit of pre-existing material, reflecting the tolerance of plants to salt stress.

The effect of NPK fertilization doses on absolute and relative growth rates of sunflower CD was only significant in the absolute growth rate, in the period from 30 to 60 DAS - TCADC1 (Table 10); however, there was no significant effect for the factor fertilization doses with chicken manure biochar and interaction between factors (D x B) for any analyzed variable.

Table 10 - Summary of the 'F' test for absolute and relative growth rate of sunflower stem diameter in the evaluation periods carried out between 30-60 (TCADC1 and TCRDC1) and 60-90 (TCADC2 and TCRDC2) days after sowing (DAS) as a function of NPK and chicken manure biochar doses

Source of Variation	F Test			
	TCADC1	TCADC21	TCRDC1	TCRDC21
Doses (D)	**	ns	ns	ns
Linear Reg.	*	-	-	-
Quadratic Reg.	*	-	-	-
Biochar (B)	ns	ns	ns	ns
D x B	ns	ns	ns	ns
CV (%)	30,61	37,85	33,06	35,56

(**), (*); (ns) significant at ($p \leq 0.01$) and ($p \leq 0.05$) and not significant, respectively
1statistical analysis performed after data transformation to $\sqrt{X}$

According to the regression studies for the TCAD1 as a function of the NPK doses (Figure 11) there was a quadratic model adjustment of the data, with the highest values obtained with the dose of 60% of the recommendation (0.1902 mm day-1) and from this point on there was a reduction of 8.41%. It can also be seen that the increase provided with the increase of NPK levels between the doses of 0 and 60% of the recommendation was 18.93%. As the CD at 30, 60 and 90 (Figures 5 and 6) a similar tendency was observed for the TCRDC1, in which the fertilizer recommendation was higher than the crop requirement, having, on average, the maximum point at a dose of 50% (50:150:75 mg kg-1) of the NPK recommendation. Prado & Leal (2006) observed reductions in sunflower growth with the omission of N, P and K. According to Furtado et al. (2014) potassium fertilization promotes increment in sunflower TCADC up to a dose of 100 mg kg-1 of soil in the evaluation period from 35 to 60 DAS. Maia Junior et al. (2013) found, for the same cultivar, an average TCADC of 0.0974 mm day-1 in the evaluation period from 20 to 80 DAS.

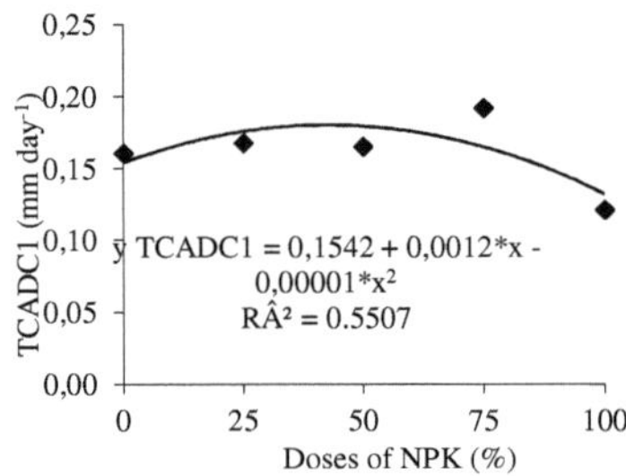

Figure 11 - Absolute growth rate of stem diameter - TCADC1 (30-60 DAS) as a function of NPK doses (%)

4.3 Phytomass Production

Fertilizing sunflower with NPK (D) had a significant effect on all variables related to phytomass production; however, the doses of chicken litter biochar (B) and the interaction between these factors (D x B), had no significant effect (Table 11).

Table 11 - Summary of the 'F' test for leaf dry phytomass (FSF), stem dry phytomass (FSC), chapter dry phytomass (FSCap) and aboveground dry phytomass (FSPA) as a function of NPK and poultry litter biochar doses

Source of Variation	F Test			
	FSF	FSC	FSCap	FSPA
Doses (D)	**	**	**	**
Linear Reg.	ns	*	*	ns
Quadratic Reg.	**	**	**	**
Biochar (B)	ns	ns	ns	ns
D x B	ns	ns	ns	ns
CV (%)	34,19	38,10	33,02	33,02

(**), (*); (ns) significant at ($p \leq 0.01$) and ($p \leq 0.05$) and not significant respectively

Although the doses of biochar limited sunflower growth, as evidenced by reductions in AP, AD, NF and AF, phytomass production remained constant within the levels of biochar fertilization; however, Coomer et al. (2012) found positive effects of fertilization with poultry litter biochar on cotton dry matter production. Suppadit et al. (2012) observed an increase in soybean biomass production with increasing doses of poultry litter biochar.

The highest values of leaf dry phytomass (FSF), stem dry phytomass (FSC), chapter dry phytomass (FSCap) and aboveground dry phytomass (FSPA) were 7,04; 8,42;5,51 and 21,09 g/plant, respectively with the doses of 45,86%; 36,28%; 37,10% and 41,65% of the NPK recommendation (Figures 12A, B, C and D); it was also observed that the fertilization with 100% (100:300:150 mg kg-1) of the recommendation promoted reduction of FSF, FSC, FSCap and FSPA, respectively of 31.73%; 43.32%; 35.98% and 32.26% in reference to the dose of maximum efficiency. These results again show that the adopted recommendation (NOVAIS et al., 1991) for the development conditions of the research was higher than the nutritional needs of sunflower. Moreover, the highest accumulation of sunflower dry matter was near the dose of 50% of the recommendation, the same trend observed for AP, AD, NF and AF (Figures 4, 5, 6, 7 and 8). According to Banerjee et al. (2014) the increase in phytomass promoted by higher levels of NPK fertilization resulted from the increase in the growth rate of the crop due to the higher production of photo-assimilates and leaf area index. According to these authors, the combination of 100:50:50 kg ha-1 of NPK promotes greater accumulation of sunflower dry matter. Abbadi et al. (2008) found an increase in FSF, FSC and FSCap with the increase of nitrogen fertilization, whose highest values were observed with a dose of 4 g/vaso of N. Salih (2013) found that the combination of 80 kg ha-1 of N and P promotes greater sunflower dry matter accumulation. For Bakhtet al. (2010) the combination of 150:100:100 kg ha-1 of NPK promotes increased sunflower dry matter accumulation.

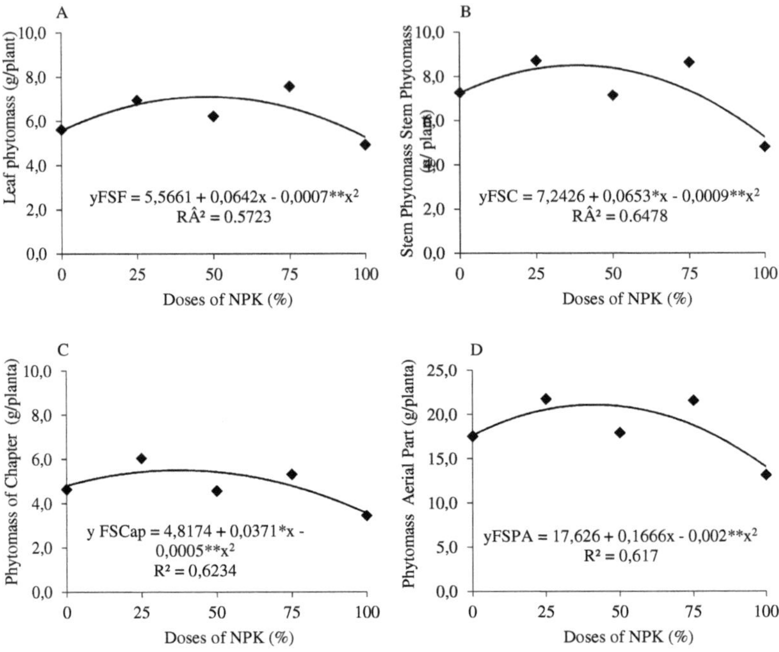

Figure 12 - Leaf dry phytomass - FSF (A), stem dry phytomass - FSC (B), chapter dry phytomass - FSCap (C) and aboveground dry phytomass - FSPA (D) of sunflower, as a function of NPK doses - %.

4.4 Production variables

The application of increasing doses of NPK in the experimental units had significant effect on sunflower yield components (Table 12); corroborating, Jahangir et al. (2006) found significant effect of N and P fertilization on number of achenes per chapter, mass of hundred achenes and grain yield; similarly, Oliveira et al. (2014) observed significant effect of nitrogen fertilization on chapter diameter (DCap) and number of achenes per chapter of sunflower.

Table 12 - Summary of the 'F' test for chapter diameter (DCap), number of achenes per chapter (NAC), number of viable achenes (NAV), percentage of non-viable achenes (PANV), mass of 100 achenes (M100A) and production of achenes per plant (PROD) as a function of doses of NPK and poultry litter biochar

Source of Variation	F Test					
	DCap	NAC	NAV	PANV1	M100A	PROD
Doses (D)	**	**	**	**	*	**
Linear Reg.	ns	**	ns	**	ns	ns
Quadratic Reg.	**	**	**	ns	*	**
Biochar (B)	**	**	**	ns	ns	**
Linear Reg.	**	**	**	-	-	**
Quadratic Reg.	ns	ns	*	-	-	**
D x B	**	ns	ns	ns	ns	*
CV (%)	15,64	33,63	16,74	36,75	21,67	28,79

(**), (*); (ns) significant at ($p \leq 0.01$) and ($p \leq 0.05$) and not significant respectively
1statistical analysis performed after data transformation to $\sqrt{X} + 1$

Fertilization with biochar significantly affected all sunflower yield variables except the percentage of non-viable achenes (PANV) and mass of one hundred achenes (M100A). According to Suppadit et al. (2012) poultry litter biochar had a significant effect on soybean yield components.

There was also a significant effect of the interaction between the D x B factors for chapter diameter (DCap) and production per plant (PROD) (Table 12), whose values of this interaction are presented in Figure 13, according to the regression equations (Table 13).

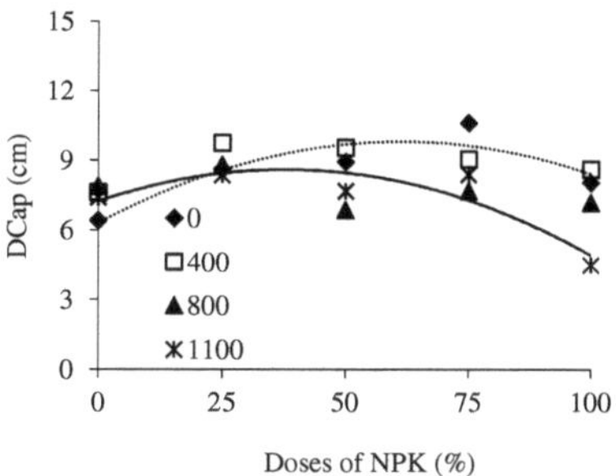

Figure 13 - Sunflower chapter diameter (CDap) as a function of NPK - % and chicken manure biochar - g/vessel

Table 13 - Regression equations for sunflower chapter diameter (CDap) as a function of NPK and chicken manure biochar doses

Chapter diameter (cm)			
Biochar (kg ha-1)	Equation	R2	Average
0	Y = 6.3268+ 0.1139**x - 0.0009**x2	0,8156	8,54
1100	Y = 7.2476+ 0.0721**x - 0.001**x2	0,7990	7,28

(**), (*) significant at ($p \leq 0.01$) and ($p \leq 0.05$); Y = chapter diameter (cm); x = NPK doses (%)

An increase in WCD was observed until a dose of 63.28% (9.93 cm) and 36.05% (8.55 mm) of the NPK recommendation, respectively, for 0 and 1100 g/vessel of biochar, with the lowest values estimated for plants fertilized with 100% of the NPK recommendation.

In relation to the plants that received the dosage of 400 and 800 g/vessel of biochar it can be seen, according to regression studies (Figure 13), that there was no significant effect on DWC, reaching average values of respectively 8.93 and 7.69 cm. As for the growth variables, the DCap of sunflower also reduced significantly with the application of biochar due to the increase in soil salinity (Table 5). Nobre et al. (2010) observed a linear reduction in the diameters of the inner and outer chapters of sunflower with the irrigation of water with EC starting at 0.5 dS m-1; according to the authors, the increase in salinity causes a reduction in the availability of water for the plants and as a consequence a reduction in the diameter of the chapter.

Regarding mineral fertilization, a similar tendency to growth variables was observed, and on average, the dose of 50% (N:P:K, 50:150:75 mg kg-1) of the recommendation promoted the highest results. According to Jahangir et al. (2006) the combination of 120 kg ha-1 of N and 75 kg ha-1 of P2O5 promotes larger DCap of sunflower, obtaining a maximum diameter of 13.3 cm. Siddiqui et al. (2009) found larger DCap with 90:45:45 kg ha-1 NPK fertilization. Ali & Noorka (2013) observed maximum DCap of 17.87 cm with the combination of 135 kg ha-1 [N] and 75 kg ha-1 [P].

According to the regression equation (Figure 14) there was a quadratic effect for the number of achenes per chapter (NAC) as a function of the NPK dose. The highest values were observed with a dose of 51.61% (430 achenes) of the recommendation, which promoted an increase of 32.43% in relation to the lowest dose. It can also be seen that from this point on, there was a reduction in NAC of 28.52% in relation to the plants that received 100% of the NPK recommendation. These results are related to the higher DCap obtained near the dose of 50% of the NPK recommendation (Figure 13). Larger buds favor greater flower formation and, in return, also a greater number of achenes. According to Zagonel & Mundstock (1991) this variable reflects the nutritional availability in the floral differentiation phase, which determines the potential number of flowers and, consequently, the number of achenes. According to Wendt et al. (2005) the size of the receptacle is a relevant component for grain yield because it reflects the potential number of achenes. Corroborating this, Pivetta et al. (2012) verified a positive and significant correlation between the number of grains per chapter and the diameter of the sunflower chapter, denoting that the increase in one variable will lead to an increase in the other. According to Jahangir et al. (2006) the combination of 120 kg ha-1 N and 75 kg ha-1 P2O5 promoted higher NAC of sunflower obtaining, on average, 341 achenes. Centeno et al. (2014) verified a linear increase in the total number of achenes with increasing doses of N, obtaining the highest results with a dose of 150 mg kg-1.

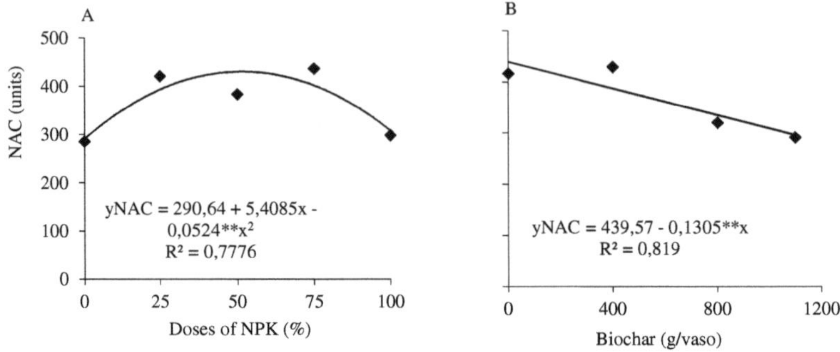

Figure 14 - Number of achenes per chapter (NAC) as a function of NPK - % (A) and chicken litter biochar - g/vessel (B)

The results for NAC as a function of the doses of biochar were best fitted to the linear model (Figure 14B). According to the regression equation, there was a reduction of 35.66% when comparing the doses 0 and 1100 g/vessel. The NAC varied from 296 to 440 units; however, these values are lower than those reported by Sachs et al. (2006) who found a variation of 688 to 818 units for Embrapa 122/V2000. The increase in soil salinity, resulting from the increase in biochar doses, also affected the NAC of sunflower. According to Rhoades et al. (2000) the effects of salt stress are manifested in both crop development and production, mainly reducing fruit population and development, symptoms similar to those of water stress. Centeno et al. (2014) found a linear reduction in the total number of sunflower achenes with increasing irrigation water EC.

According to the regression equation (Figure 15A) there was a quadratic effect on the number of viable achenes of sunflower (NAV) as a function of NPK doses, indicating that increasing the dose up to 54.48% (374 units) of the NPK recommendation promoted an increase of 41.95% in relation to the lowest dose. It can also be observed that from this point on, there was a reduction of 29.25% in relation to the plants that received 100% (N:P:K, 100:300:150 g/vaseo) of the recommendation.

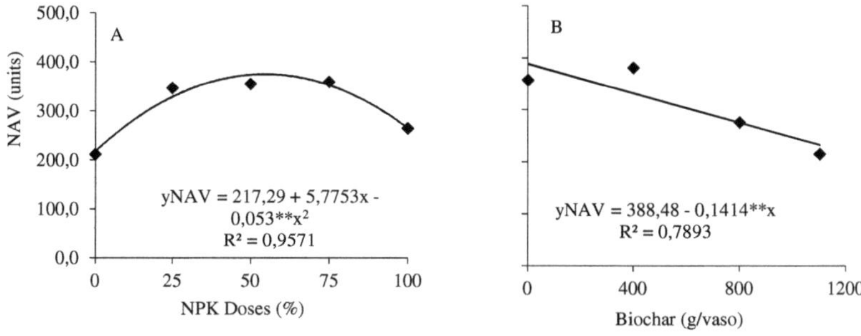

Figure 15 - Number of viable achenes (NAV) as a function of NPK doses - % (A) and chicken litter biochar - g/vessel (B)

As for TSD (Figure 13) and NAC (Figure 14), fertilization around 50% of the NPK recommendation promoted greater NAV, which is justified by the adequate availability of nutrients, especially in the flowering phase (R1 stage). According to Lobo & Grassi Filho (2007), the diameter of the chapter is closely related to the potential number of achenes; in this sense, increasing the DCap will increase the NAC and, consequently, the NAV, an essential component of productivity. Pivetta et al. (2012) found a positive and significant correlation between DCap and the percentage of normal grains.

Fertilization with biochar promoted a linear reduction in the NAV of sunflower (Figure 15B), with the lowest values observed with the dose of 1100 g/vessel (233 units), which corresponded to a reduction of 40.03% in reference to plants that were not fertilized with biochar (388 units). Again, the effect of increased soil salinity due to biochar fertilization is observed. The increase in excess soluble salts in the soil promotes increased osmotic potential and reduced water potential, reducing the photosynthetic rate and, consequently, the production of photoassimilates for the reproductive organs. Santos Junior et al. (2011a) verified a linear reduction of NAV with increasing irrigation water EC and obtained an average production of 300 units at a level of 4.3 dS m-1.

Corroborating Centeno et al. (2014) found a linear reduction in the percentage of viable achenes with increasing EC of irrigation water.

The results of the percentage of non-viable achenes (NAPV) as a function of the NPK dose were best fitted to a decreasing quadratic model, indicating that the increase in fertilization up to a dose of 73.5% of the recommendation promoted a reduction in the NAPV of 55.23% in relation to the lowest dose and from this point on there was an increase of 13.78% in relation to the dose of 100% of the recommendation (Figure 16A). These results are justified by the greater nutritional availability near the dose of 50% of the recommendation. The NAP varied from 10.51% to 23.47%; however, these values are lower than those reported by Santos Junior et al. (2011b) who obtained a variation of 21.70% to 35.81%. According to Garcia et al. (2007), progressive salt stress may induce ionic toxicity and/or nutritional imbalance due to the excessive accumulation of ions in the cellular protoplasm, which may cause cellular injury through oxidative stress, which may result in a decrease in the number of achene fruit, mass of achene fruit and, consequently, a greater number of shocked grains. It was also observed that the NWFP showed a tendency opposite to the behavior of the NAV (Figure 15A), which shows coherence between the results, i.e., as there is an increase in the NAV there is a reduction in the NWFP, which is justified since the maximum estimated doses were close.

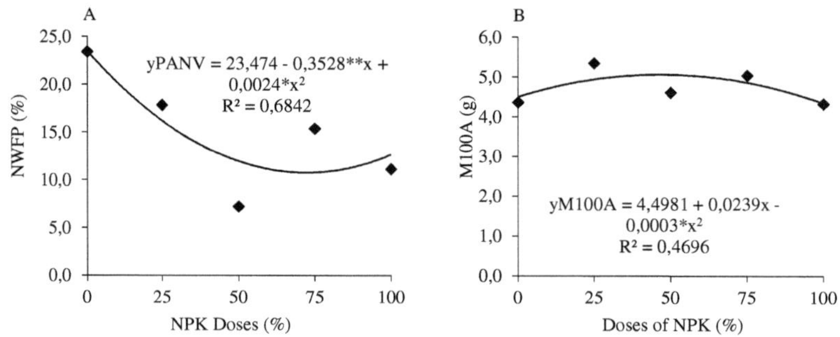

Figure 16 - Percentage of non-viable achenes (PANV) (A) and mass of one hundred achenes (M100A) (B) as a function of NPK - % doses

In relation to the mass of one hundred achenes (M100A) (Figure 16B), there was a better fit of the data to the quadratic model, obtaining a maximum point at a dose of 39.83% of the NPK recommendation, corresponding to an increase of 9.49% in relation to the lowest dose. M100A ranged from 3.88 to 4.97 g; however, these values are lower than those reported by Jahangir et al. (2006), who observed variation from 5.01 to 5.73 g, and obtained maximum values with combination of 120 kg ha-1 N and 75 kg ha-1 P2O5. Nobre et al. (2011) verified a linear effect of nitrogen fertilization on this variable, and plants submitted to a dose of 125 mg kg-1 produced M100A equal to 4.87 g. For Banerjee et al. (2014) the combination of 100:50:50 kg ha-1 of NPK promoted higher M100A (5.63 g). Corroborating, Ali & Noorka (2013) found that the combination of 135 and 75 kg ha-1 of NP promotes higher mass of thousand achenes (51.93 g). These results are similar to those obtained for DCap (Figure 13), NAC (Figure 14A) and NAV (Figure 15A). According to Amorim et al. (2008) there are significant effects between grain yield, chapter diameter, percentage of normal grains and mass of thousand grains.

The interaction between factors (D x B) for the production of achenes per plant (PROD) can be seen in Figure 17. According to the regression equations (Table 14), there was a quadratic effect in plants fertilized with 0, 400 and 1100 g/vessel of biochar, with an increase in the yield up to 59.17% (21.29 g/plant), 49.60% (23.91 g/plant) and 48.06% (15.68 g/plant) of the recommendation, respectively. The lowest values were estimated for plants fertilized with 100% (100:300:150 mg kg-1) of the NPK recommendation.

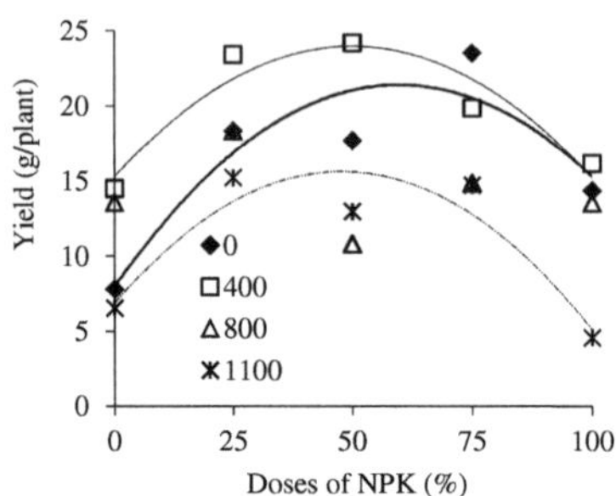

Figure 17 - Production of achenes per plant (PROD) of sunflower as a function of doses of NPK - % and poultry litter biochar - g/vessel

Table 14 - Regression equations for sunflower achene production per plant as a function of NPK and chicken litter biochar doses

	Production per Plant (g)		
Biochar (g/vessel)	Equation	R2	Average
0	Y = 7.9742+ 0.4497*x - 0.0038**x2	0,8258	16,34
400	Y = 15.308+ 0.3472x - 0.0035**x2	0,8939	19,62
1100	Y = 6.9056+ 0.3653x - 0.0038**x2	0,8583	10,80

(**), (*) significant at ($p \leq 0.01$) and ($p \leq 0.05$); Y = achene production per plant (g/plant); x = NPK doses (%)

For the plants that received the dose of 800 g/vessel of biochar, according to the regression studies (Figure 17) there was no significant effect on the yield, with an average value of 14.21 g/plant. As for the other production components, it is observed that the highest yield was near the dose of 50% of the recommendation, a fact related to M100A (Figure 16B). According to Yasin et al. (2014), the mass of achene expresses the degree of grain development and plays a key role in assessing the potential yield of a crop. According to Banerjee et al. (2014) the combination of 100:50:50 kg ha-1 of NPK promoted higher sunflower achene yield (1,465.15 kg ha-1). For Jahangir et al. (2006) the combination of 120 kg ha-1 of N and 75 kg ha-1 of P2O5 promoted higher aquenium yield (2,001 kg ha-1); however, Ali & Noorka (2013) observed better sunflower yield with the combination of 135 and 75 kg ha-1 of N and P (2,584 kg ha-1).

It can also be seen that, although fertilization with biochar promoted a reduction in production components, as evidenced by the values of FCD (Figure 13), NAC (Figure 14B) and NAV (Figure 15B), the dose of 400 g/vessel (19.62 g/plant) promoted an increase in yield of 16.72% compared to plants that were not fertilized with biochar (16.34 g/plant). According to Suppadit et al. (2012) the application of 98.6 g/vessel of chicken litter biochar promoted higher soybean seed yield. According to Tagoe et al. (2008) the combination of 11.4 g/vessel of poultry litter biochar with 410 mg kg-1 of K promoted increases in soybean and cowpea yields of 53 and 185%, respectively, relative to the control (no fertilization).

It was also observed that the dose of 1100 g/vine (10.80 g/plant) promoted a reduction of 44.95% in the yield in relation to the dose of 400 g/vine, probably due to the excess of salts in the soil. Salt stress promotes metabolic disorders in plants, affecting their physiology, promoting a reduction in the absorption of water and nutrients from the soil, mainly due to the reduction of leaf area, implying a smaller photosynthesizing surface and, consequently, lower crop yield (LIMA et al. 2014). According to Ferreira Neto et al. (2007), in salt stress conditions the content of NaCl in the soil solution is high, which can lead to losses in the absorption of mineral nutrients, especially NO3-, K+ and Ca2+, promoting a reduction in the growth, development and yield of the crop. Nobre et al. (2011) observed a linear reduction in yield with increasing EC of irrigation water, obtaining 17.4 g/plant when the plants were irrigated with higher levels of EC. Similar results were also observed by Centeno et al. (2014).

CONCLUSIONS

1. The fertilization with chicken manure, starting at 400 g/vase, promoted an increase in the salinity of the soil in the experimental units;

2. Fertilization with 50% of the NPK recommendation (50:150:75 mg kg-1) promoted the highest values of sunflower growth, phytomass production, production components, and yield when compared to the other treatments at NPK levels;

3. Fertilization with chicken litter biochar promoted reduced growth and yield components of sunflower;

4. The fertilization with 400 g/vaseo of chicken manure biochar promoted an increase of 16.72% in the production of achenes per plant (PROD);

5. The combination of 50:150:75 mg kg-1 of NPK and 400 g/vessel of biochar promotes higher PROD of sunflower (23.91 g/plant).

REFERENCES

ABBADI, J.; GERENDÁS, J.; SATTELMACHER, B. Effects of nitrogen supply on growth, yield and yield components of safflower and sunflower. Plant and Soil, v.306, n.1-2, p.167-180, 2008.

ADEBAYO, A.G.; AKINTOYE, H.A.; OLATUNJI, M.T.; SHOKALU A.O.; AINA O.O. Growth response and flower yield of sunflower to phosphorus fertilization in Ibadan, Southwestern Nigeria. Report and Opinion, v.2, n.4, p.29-33, 2010.

AESA - Executive Agency for Water Management of the State of Paraíba. Campina Grande: AESA, 2014.

ALI, A.; NOORKA, I.R. Nitrogen and phosphorus management strategy for better growth and yield of sunflower (*Helianthus annuus* L.) hybrid. Soil & Environment, v.32, n.1, p.44-48, 2013

AMANULLAH; KHAN, M.W. Interactive effects of potassium and phosphorus on phenology and grain yield of sunflower in Northwest Pakistan. Pedosphere, v.20, n.5, p.674-680, 2010.

AMORIM, E.P.; RAMOS, N.P.; UNGARO, M.R.G.; KIIHL, T.A. Correlations and path analysis in sunflower. Bragantia, v.67, n.2, p.307-316, 2008.

AQUINO, L.A.; SILVA, F.D.B.; BERGER, P.G. Agronomic characteristics and the nutritional status of irrigated sunflower cultivars. Brazilian Journal of Agricultural and Environmental Engineering, v.17, n.5, p.551-557, 2013.

ÁVILA, V.S.; MAZZUCO, H.; FIGUEIREDO, E.A.P. Cama de aviário: materiais, reutilização, uso como alimento e fertilizante. Concórdia, SC: Embrapa-CNPSA, 1992. 38p. (Technical Circular 16).

AYERS, R.S.; WESTCOT, D.W. A qualidade de água na agricultura. 2. ed. Campina Grande: UFPB, 1999, 153 p. (FAO, Irrigation and Drainage, 29).

BAKHT, J.; SHAFI, M.; YOUSAF, M.; SHAH, H.U. Physiology, phenology and yield of sunflower (AUTUMN) as affected by NPK fertilizer and hybrids. Pakistan Journal Botany, v.42, n.3, p.1909-1922, 2010.

BANERJEE, H.; DUTTA, S.K.; PRAMANIK, S.J.; RAY,K.; PHONGLOSA, A.; BHATTACHARYYA, K. Productivity and profitability of spring planted sunflower hybrid with nitrogen, phosphorus and potassium fertilizer. Annals of Plant and Soil Research, v.16, n.3, p.250-256, 2014.

BERNARDI, A.C.C.; SOUZA, G.B.; OKA, S.H.; RASSINI, J.B. Equivalence of methods in the extraction of potassium from the aerial part of alfalfa and tanzânia grass with water or with concentrated or diluted acid solution. Embrapa Pecuária, 2008. 5p. (Technical Communication 86).

BÍSCARO, G.A.; MACHADO, J.R.; TOSTA, M.S.; MENDONÇA, V.; SORATTO, R.P.; CARVALHO, L.A. Covering nitrogen fertilization in irrigated sunflower under Cassilândia-MS conditions. Ciência e Agrotecnologia, v.32, n.5, p.1366-1373, 2008.

BLUM, L.E.B.; AMARANTE, C.V.T.; GÜTTLER, G; MACEDO, A.F.; KOTHE, D.M.; SIMMLER, A.O.; PRADO, G.; GUIMARÃES, L. Strawberry and cucumber production in soil with incorporation of poultry litter and pine bark. Horticultura Brasileira, v.21, n.4, p.627-631, 2003.

BOSCO, M.R.O.; OLIVEIRA, A.B.; HERNANDEZ, F.F.F.; LACERDA, C.F. Influence of salt stress on the mineral composition of eggplant. Revista Ciência Agronômica, v.40, n.2, p. 157-164, 2009.

BRASIL. Rules for seed analysis. Ministry of Agriculture, Livestock and Supply. Secretary of Agricultural Defense. Brasília: Mapa/ACS, 2009. 399p.

CAMPOS, V.B.; CHAVES, L.H.G.; GUERRA, H.O.C. GUEDES FILHO, D.H.; SANTOS JUNIOR, J.A. Adubação NPK e conteúdo de água disponível no solo na cultura do girassol para fins ornamentais. In: Congresso Norte e Nordeste de Pesquisa e Inovação, 7. 2012, Palmas. Annals Palmas, 2012. <

http://propi.ifto.edu.br/ocs/index.php/connepi/vii/paper/viewFile/28/1328>. 15 Nov. 2014.

CANTARUTTI, R.C.; NEVES, J.C. L. Fertilidade do Solo. 1 ed. Viçosa: SBCS, 2007. p. 91-132.

CARVALHO, E.R.; REZENDE, P.M.; ANDRADE, M.J.B.; PASSOS, A.M.A.; OLIVEIRA, J.A. Mineral fertilizer and organic residue on soybean agronomic characteristics and soil nutrients. Revista Ciência Agronômica, v.42, n.4, p.930-939, 2011.

CASTRO, C.; FARIAS, J.R.B. Ecofisiologia do Girassol. In: LEITE, R.M.V.B.C.; BRIGHENTI, A.M.; CASTRO, C. (ed.). Girassol no Brasil. Londrina: Embrapa Soja, 2005. p.163-218.

CASTRO, C.; OLIVEIRA, F.A. Nutrição e Adubação do Girassol. In: LEITE, R.M.V.B.C.; BRIGHENTI, A.M.; CASTRO, C. (ed.). Girassol no Brasil. Londrina: Embrapa Soja, 2005. p.317-373.

CENTENO, C.R.M.; SANTOS, J.B.; XAVIER, D.A.; AZEVEDO, C.A.V.; GHEYI, H.R. Production components of sunflower Embrapa 122-V2000 under water salinity and nitrogen fertilization. Brazilian Journal of Agricultural and Environmental Engineering, v.18, (Supplement), p.S39-S45, 2014.

CHAN, Y. K.; VAN ZWIETEN, L.; MESZAROS, I.; DOWNIE, A.; JOSEPH, S. Using poultry litter biochars as soil amendments. Soil, Land Care & Environmental Research, v.46, n.5, p.437-444, 2008.

CHAVES, L.H.G.; GUERRA, H.O.C.; CAMPOS, V.B.; PEREIRA, W.E.; RIBEIRO, P.H.P. Biometry and water consumption of sunflower as affected by NPK fertilizer and available soil water content under semiarid Brazilian conditions. Agricultural Sciences, v.5, n.8, p.668-676, 2014.

COELHO, M.A.; SONCIN, N.B. Geografia do Brasil. São Paulo: Moderna. 1982. 368p.

CONAB - Companhia Nacional de Abastecimento. Acompanhamento da safra brasileira de grãos 2012/13 - Décimo Segundo Levantamento - September/ 2013. < http://www.conab.gov.br/OlalaCMS/uploads/arquivos/13_10_16_14_32_01_ boletim_portugues_-_setembro_2013.pdf>. 08 Sep. 2014.

CONNOR, J.D.; HALL, A. J. Sunflower physiology. In: SCHNEIDER, A. A. (ed.). Sunflower technology and production. Madison: ASA/CSSA/SSSA, 1997. p. 113-181. (Series of monographs, 35).

COOMER, T.D.; OOSTERHUIS, D.M.; LONGER, D.E.; LOKA, D.A. Effect of poultry litter biochar on early-season cotton growth. Summaries of Arkansas Cotton Research, p.89-94, 2012. (Research Series 610).

CORRÊA, J. C.; MIELE, M.A cama de aves e os aspectos agronômicos, ambientais e econômicos. In: PALHARES, J.C. P.; KUNZ, A. (ed.). Manejo ambiental na avicultura. Concórdia: Embrapa Suínos e Aves, 2011. p. 125-152. (Documentos, 149).

COSTA, A.M.; BORGES, E.N.; SILVA, A.A.; NOLLA, A.; GUIMARÃES, E.C. Potential for physical recovery of a red latosol under degraded pasture influenced by the application of chicken litter. Ciência agrotecnológica. v.33, (Edição especial), p.1991-1998, 2009.

CZERNIK, S.; BRIDGWATER, A. V. Overview of applications of biomass fast pyrolysis oil. Energy and Fuels, v.18, n.2, p.590-98, 2004.

DALL'AGNOL, A.; VIEIRA, O.V.; LEITE, R.M.V.B.C. Origem e histórico do girassol. In: LEITE, R.M.V.B.C.; BRIGHENTI, A.M.; CASTRO, C. (ed.). Girassol no Brasil. Londrina: Embrapa Soja, 2005. p.1-14.

DECHEN, A. R.; NACHTIGALL, G. R. Elementos requeridos à nutrição de plantas. In: NOVAIS, R.F.; ALVAREZ, V.V.H.; BARROS, N. F.; FONTES, R.L. F.; ANTARUTTI, R.B; NEVES, J.C.L. (ed.). Fertilidade do solo. Viçosa: SBCS/UFV, 2007. p.91-132.

DIKINYA, O.; MUFWANZALA, N. Chicken manure-enhanced soil fertility and productivity: Effects of application rates. Journal of Soil Science and Environmental Management, v.1, n.3, p.46-54, 2010.

EMBRAPA - Brazilian Agricultural Research Corporation. Sunflower Embrapa 122/V-2000. Londrina, 2006. (Folder n. 04/2006).

EMBRAPA - Empresa Brasileira de Pesquisa Agropecuária. Manual de análises químicas de solos, plantas e fertilizantes. 2 ed. Brasília: Embrapa Informação Tecnológica, 2009. 627p

EMBRAPA - Empresa Brasileira de Pesquisa Agropecuária. Manual de métodos de análise de solo. 2 ed. Rio de Janeiro: Embrapa Solos, 2011. 230p.

FAGUNDES, J.D.; SANTIAGO, G.; MELLO, A.M.; BELLÉ, R.A.; STRECK, N.A. Growth, development and retardation of leaf senescence in pot sunflower (*Helianthus annuus* L.): nitrogen sources and doses. Ciência Rural, v.37, n.4, 2007.

FALCÃO, N.P. S.; CARVALHO, E.J. M.; COMERFORD, N. Avaliação da fertilidade de solos antropogênicos da Amazônia Central. In: Congresso da Sociedade de Arqueologia Brasileira, 11. Working Group: Archaeological Black Earth in the Amazon: State of the Art. Rio de Janeiro, 2001. 2p.

FAO (Rome). FAOSTAT. <http://faostat3.fao.org/faostat-gateway/go/to/download/Q/QC/S>. 25 Oct. 2013.

FERREIRA NETO, M.; GHEYI, H.R.; FERNANDES, P.D.; HOLANDA, J. S.; BLANCO, F.F. Leaf emission, ionic relations and production of coconut irrigated with saline water. Ciência Rural, v.37, n.6, p.1675-1681, 2007.

FERREIRA, D.F. Sisvar: a computer statistical analysis system. Ciência e Agrotecnologia, v.35, n.6, p.1039-1042, 2011.

FLOWERS, T. J. Improving crop salt tolerance. Journal of Experimental Botany, v.55, n.396, p.307-319, 2004.

FOX, T. A.; CHAPMAN, L. Engineering geo-engineering. Meteorological Applications, v.18, n.1, p.1-8, 2011.

FURTADO, G.F.; XAVIER, D.A.; ANDRADE, E.M.G.; SOARES, L.A.A.; SOUSA, J.R.M.; LIMA, G.S. Growth rates of sunflower grown under different levels of water replacement and doses of K. In: UMS: Soil Use and Management Congress, 6. 2014, Recife. Proceedings Recife, p.257-260, 2014.

GARCIA, G. O.; NAZÁRIO, A.A.; MORAES, W.B.; GONÇALVES, I.Z.; MADALÃO, J.C. Responses of bean genotypes to salinity. Engenharia na Agricultura, v.18, n.4, 2010.

GIANELLO, C.; ERNANI, P.R. Rendimento de matéria seca de milho e alterações na composição química do solo pela incorporação de quantidades crescentes de cama de frangos, em casa de vegetação. Revista Brasileira de Ciência do Solo, v.7, n.3, p.285-290, 1983.

GLASER, B.; LEHMANN, J.; ZECH, W. Ameliorating physical and chemical properties of highly weathered soil in the tropics with charcoal - a review. Biology and Fertility of Soils, v.35, n.4, p.219-230, 2002.

HOLANDA, J.S.; AMORIM, J.R.A.; NETO, M.F.; HOLANDA, A.C. Qualidade da água para irrigação. In: GHEYI, H.R.; DIAS, N.S.; LACERDA, C.F. (ed.). Manejo da salinidade na agricultura: Estudos básicos e aplicadas. Fortaleza: INCT Sal, 2010. 472p.

IOANNIDOU, O.; ZABANIOTOU, A. Agricultural residues as precursors for activated carbon production - a review. Renewable Sustainable Energy Reviews, v.11, n.9, p.1966-2005, 2007.

JAHANGIR, A.A.; MONDAL, R.K.; NADA, K.; AFROZE, S.R.; HAKIM, M.A. Response of nitrogen and phosphorus fertilizer and plant spacing on growth and yield contributing character of sunflower. Bangladesh Journal of Science and Industrial Research, v.41, n.1-2, p.33-40, 2006.

JIEN, S-H.; WANG, C-S. Effects of biochar on soil properties and erosion potential in a highly weathered soil. CATENA, s.n., v.110, p.225-233, 2013.

KOOKANA, R.S.; SARMAH, A.K.; VAN ZWIETEN L. Biochar application to soil: agronomic and environmental benefits and unintended consequences. Advances in Agronomy, v.112, s.n., p.103-143, 2011.

LEHMANN, J. Bio-energy in the black. Frontiers in Ecology and the Environment, v.5, n.7, p.381-387, 2007.

LEHMANN, J., GAUNT, J., RONDON, M. Bio-char sequestration in terrestrial ecosystems - a review. Mitigation and Adaptation Strategies for Global Change, v.11, n.2, p.403-427, 2006.

LEHMANN, J.; SILVA JR., J.P.; STEINER, C.; NEHLS, T.; ZECH, W.; GLASER,B. Nutrient availability and leaching in an archaeological anthrosol and a ferralsol of the Central Amazon basin: Fertilizer, manure and charcoal amendments. Plant and Soil, v.249, n.2, p.343-357, 2003.

LEHMANN, J.; JOSEPH, S. Biochar for Environmental Management: Science and Technology. London: Earthscan, 2009. 448p.

LIESCH, A.M.; WEYERS, S.L.; GASKIN, J.W.; DAS, K.C. Impact of two different biochars on earth worm growth and survival. Annals of Environmental Science, v.4, n.1, p.1-9, 2010.

LIMA, A.D.; VIANA, T.V.A.; AZEVEDO, B.M.; MARINHO, A.B.; DUARTE, J.M.L. Adubação borácica na cultura do girassol. Revista Agro@mbiente On-line, v.7, n.3, p.269-276, 2013.

LIMA, G.S.; NOBRE, R.G.; GHEYI, H.R.; SOARES, L.A.A.; SILVA, A.O. Growth and production components of castor oil plant under salt stress and nitrogen fertilization. Agricultural Engineering, v.34, n.5, p.854-866, 2014.

LIMA, I.M.; MCALOON, A. J.; BOATENG, A.A. Activated carbon from broiler litter: process description and cost of production. Biomass Bioenergy, v.32, n.6, p.568-572, 2008.

LIRA, M.A.; CARVALHO, H.W.L.; CHAGAS, M.C.M.; BRISTOT, G.; DANTAS, J.A.; LIMA, J.M.P. Avaliação das potencialidades da cultura do girassol, como alternativa de cultivo no semárido nordestino. Natal: EMPARN, 2011. 40p. (Documents 40).

LOBO, T. F.; GRASSI FILHO, H. Levels of sewage sludge in sunflower productivity. Revista de la Ciencia del Suelo y Nutrición Vegetal, v.7, n.3, p.16-25, 2007.

MADARI, B.E.; COSTA, A.R.; CASTRO, L.M.; SANTOS, J.L.S.; BENITES, V. M.; ROCHA, A. O.; MACHADO, P.L.O.A. Charcoal as soil conditioner for upland rice (Primavera cultivar): a prospective study. Santo Antônio de Goiás: Embrapa Arroz e Feijão, 2006. 6p. (Technical communication, 125).

MADARI, B.E.; MAIA, C.M.B.F.; NOVOTNY, E.H. Context and importance of biochar research. Pesquisa agropecuária brasileira, v.47, n.5, p.i-ii, 2012.

MAIA JUNIOR, S.O.; ANDRADE, J.R.; ARAÚJO, D.L.; SOUSA, J.S.; MEDEIROS, I.F.S. Growth rates of sunflower cultivars under different water regimes. Green Journal of Agroecology and Sustainable Development, v.8, n.3, p.150-155, 2013.

MALDANER, I.C.; HELDWEIN, A.B.; LOOSE, L.H.; LUCAS, D.D.P.; GUSE, F.I.; BORTOLUZZI, M.P. Models for non-destructive determination of leaf area in sunflower. Sci. Rural, v.39, n.5, p.1356-1361, 2009.

MATOVIC, D. Biochar as aviable carbono sequestration option: Global and canadian perspective. Energy, v.36, n.4, p.2011-2016, 2011.

MELO, A.S.; SILVA JÚNIOR, C.D.; FERNANDES, P.D.; SOBRAL, L.F.; BRITO, M.E.B.; DANTAS, J.D.M. Alterações das características fisiológicas da bananeira sob condições de fertirrigação. Revista Ciência Rural, v.39, n.3, 2009.

MEURER, E.J. Potassium. In: FERNANDES, M.S. (ed.). Nutrição mineral de plantas. Viçosa: SBCS, 2006. p.281-298.

MIELE, A.; MILAN, P.A. Composição mineral de cama de aviário de frangos de corte e sua utilização na adubação de vinhedos. Pesquisa Agropecuária Brasileira, v.18, n.7, p.729-733, 1983.

MORAIS, F.A.; GURGEL, M.T.; OLIVEIRA, F.H.T.; MOTA, A.F. Influence of irrigation with saline water on sunflower culture. Revista Ciência Agronômica, v. 42, n. 2, p. 327-336, 2011.

MUKHERJEE, A.; LAL, R. Biochar impacts on soil physical properties and greenhouse gas emissions. Agronomy, v.3, n.2, p.313-339, 2013.

NASIM, W.; AHMAD, A.; WAJID, A.; AKHTAR, J.; MUHAMMAD, D. Nitrogen effects on growth and development of sunflower hybrids under agroclimatic conditions of Multan. Pakistan Journal of Botany, v.43, n.4, p. 2083-2092, 2011.

NOBRE, R.G.; GHEYI, H.R.; CORREIA, K. G.; SOARES, F. A. L.; ANDRADE, L. O. Sunflower growth and flowering under salt stress and nitrogen fertilization. Revista Ciência Agronômica, v.41, n.3, p. 358-365, 2010.

NOBRE, R.G; GHEYI, H.R.; SOARES, F.A.L.; CARDOSO, J.A.F. Sunflower production under salt stress and nitrogen fertilization. Revista Brasileira de Ciência do Solo, v.35, n.3, p.929-937, 2011.

NOBRE, R.G; LIMA, G.S.; GHEYI, H.R.; SOARES, L.A.A.; SILVA, A.O. Crescimento, consumo e eficiência do uso da água pela mamoneira sob estresse salino e nitrogen. Revista Caatinga, v.27, n.2, p.148-158, 2014.

NOVAIS, R.F.; NEVES, J.C.L.; BARROS, N.F. Ensaio em ambiente controlado. In: OLIVEIRA A.J. (ed.). Métodos de pesquisa em fertilidade do solo. Brasília: EMBRAPA-SEA. p.189-253, 1991.

OLIVEIRA, C.R.; OLIVEIRA, J.L.; BARBOSA, F.R.; DARIO, A.S.; MOURA, S.G.; BARROS, H.B. Effect of nitrogen cover on sunflower productivity in Tocantins State. Científica, v.42, n.3, p.233-241, 2014.

OLIVEIRA, F.A.; MEDEIROS, J.F.; OLIVEIRA, M.K.T.; SOUZA, A.A.T. FERREIRA, J.A.; SOUZA, M.S. Interaction between salinity and biostimulant in cowpea bean culture. Revista Brasileira de Engenharia Agrícola e Ambiental, v.17, n.5, p.465-471, 2013.

OSMAN, E.B.A.; AWED, M.M.M. Response of sunflower (*Helianthus* annuusL.) to phosphorus and nitrogen fertilization under different plant spacing at New Valley. Assiut University Bulletin for Environmental Researches, v.13, n.1, p.11-19, 2010.

PESSOA, L.G.M.; OLIVEIRA, E.E.M.; FREIRE, M.B.G.; FREIRE, F.J.; MIRANDA, M.A.; SANTOS, R.L. Chemical composition and salinity of leachate in two soils cultivated with onion irrigated with saline water. Revista Brasileira de Ciências Agrárias,v.5, n.3, p.406-412, 2010.

PETTER, F.A.; MADARI, B.E.; SILVA, M.A.S.; CARNEIRO, M.A.C.; CARVALHO, M.T.M.; MARIMON JÚNIOR, B.H.; PACHECO, L. Soil fertility and upland rice yield after biochar application in the Cerrado. Pesquisa agropecuária brasileira, v.47, n.5, p.699-706, 2012.

PIVETTA, L.G.; GUIMARÃES, V.F.; FIOREZE, S.L.; PIVETTA, L.A.; CASTOLDI, G. Evaluation of sunflower hybrids and relationship between productive and qualitative parameters. Revista Ciência Agronômica, v.43, n.3, p.561-568, 2012.

PRADO, R.M.; LEAL, R.M. Nutritional disorders by deficiency in sunflower var. Catissol-01. Pesquisa Agropecuária Tropical, v.36, n.3, p.187-193, 2006.

RANDALL, E.L. Improved method for fat and oil analysis by a new process of extraction. Journal of the Association of Official Analytical Chemists, v.57, n.5, p.1165-1168, 1974.

REVELL, K.T. The Effect of fast pyrolysis biochar made from poultry litter on soil properties and plant growth. Blacksburg: Virginia Tech, 2011. 83p. Thesis masters.

REZENDE, E.I.P.; ANGELO, L.C.; SANTOS, S.S.; MANGRICH, A.S. Biochar and carbon sequestration. Revista Virtual Química, v.3, n.5, p.426-433, 2011.

RHOADES, J. D.; KANDIAH, A., MASHALI, A.M. Uso de águas salinas para produção agrícola. Campina Grande: UFPB, 2000. 117p. (FAO. Studies on Irrigation and Drainage, 48).

RIBEIRO, M. R.; FREIRE. F. J.; MONTENEGRO, A. A. Halomorphic soils in Brazil: Occurrence, genesis, classification, use and sustainable management. In: CURI, N.; MARQUES, J.J.; GUILHERME, L.R.G.G.; LIMA, J.M.; LOPES, A.S.; ALVAREZ, V.V.H. (ed.). Tópicos em Ciência do Solo. Viçosa: Brazilian Society of Soil Science, 2003. v.3, p.165-208.

RIBEIRO, P.H.P.; GHEYI, H.R.; SOARES, F.A.L.; UYEDA, C.A.; TEIXEIRA, M.B.; GUERRA, H.O.C. Growth rate and yield of sunflower irrigated with saline water under nitrogen doses. In: Inovagri International Meeting, 2., 2014. Fortaleza. Proceedings Fortaleza, p.3493-3502, 2014.

RICHARDS, L. A. Diagnoses and improvement of saline and alkali soils. Washington: United States Salinity Laboratory, 1954. 160p.

ROSSI, R.O. Girassol. Curitiba: Tecnoagro, 1998. 333 p.

SACHS, L.G.; PORTUGAL, A.P.; PRUDENCIO-FERREIRA, S.H.; IDA, E.I.; SACHS, P.J.D.; SACHS, J.P.D. Effect of NPK on sunflower yield and chemical components. Semina: Agricultural Sciences, v.27, n.4, p.533-546, 2006.

SACHS, L.G.; PRUDENCIO-FERREIRA, S.H.; SACHS, J.P.D.; SACHS, P.J.D.; PORTUGAL, A.P.; FELINTO, A.S. Sunflower meal: II - effect on bread quality. In: Reunião Nacional de Pesquisa de Girassol, 16 e Simpósio Nacional sobre a Cultura do Girassol, 4, 2005, Londrina. Proceedings Londrina: Embrapa Soja, p.156-159, 2005. (Documents 261).

SALIH, M.N.T. Response of sunflower (*Helianthus annuus* L.) to phosphorus and nitrogen fertilization under rainfed conditions, Blue Nile State-Sudan. Helia, v.36, n.58, p. 101-110, 2013.

SANTOS JÚNIOR, J.A. Produção de girassol em diferentes sistemas de cultivo, águas e adubação. Campina Grande: UFCG, 2010. 152p. Master's Dissertation.

SANTOS JÚNIOR, J.A.; GHEYI, H.R.; DIAS, N.S.; SOARES, F.A.L.; NOBRE, R.G. Doses of boron and wastewater in sunflower production. Revista Ciência Agronômica, v.42, n.4, p.857-864, 2011b.

SANTOS JÚNIOR, J.A.; GHEYI, H.R.; GUEDES FILHO, D.H.; DIAS, N.S.; SOARES, F.A.L. Sunflower cultivation in hydroponic system under different salinity levels. Revista Ciência Agronômica, v.42, n.4, p.842-849, 2011a.

SANTOS, G.C.G.; BERTON, R.S.; SOARES, A.G.; MARTINS, R.; DELGADO JÚNIOR, J.P. Poultry litter biochar as a source of P and its effect on soil fertility. In: Congresso Brasileiro de Ciência do Solo, 34., 2013, Florianópolis, SC. Proceedings Santa Catarina: EPAGRI, 2008.

SANTOS, J.F.; GRANGEIRO, J.I.T. Doses of chicken litter in relation to sunflower production components. Tecnologia & Ciência Agropecuária, v.7, n.2, p.15-20, 2013.

SANVONG, C.; SUPPADIT, T. The Characteristic of pelleted broiler litter biochar derived from pilot scale pyrolysis reactor and 200-liter-oil-drum kiln. Journal of Energy Technologies and Policy, v.3, n.10, p.34-38, 2013.

SCHERER, E.E. Avaliação do manco de aves e da urea como fontes de nitrogen para a cultura do milho. Agropecuária Catarinense, v.8, n.4, p.15-18, 1995.

SCHNEITER, A.A.; MILLER, J.F. Description of sunflower growth stages. Crop Science, Madison, v.21, n.6, p.901-903, 1981.

SIDDIQUI, M.H.; OAD, F.C.; ABBASI, M.K.; GANDAHI, A.W. Effect of NPK, micronutrients and N-placement on the growth and yield of sunflower. Sarhad Journal of Agriculture, v.25, n.1, 2009.

SILVA, H.L.; TREZZI. M.M.; MARCHESE, J.A.; BUZZELLO, G.; MIOTTO JR.E.; PATEL, F.; DEBASTIANI, F.; FIORESE, J. Determination of indicator species and comparison of sunflower genotypes for allelopathic potential. Planta Daninha, v.27, n.4, p.655-663, 2009.

SMIDERLE, O.J.; MOURÃO JR., M.; GIANLUPPI, D. Avaliação de cultivares de girassol em savana de Roraima. Acta Amazônica, v.35, n.3, p. 331-336, 2005.

STEINER, C.; TEIXEIRA, W. G.; LEHMANN, J.; NEHLS, T.; MACÊDO, J. L.V.; BLUM, W.E.H.; ZECH, W. Long term effects of manure, charcoal and mineral fertilization on crop production and fertility on a highly weathered Central Amazonian upland soil. Plant Soil, v.291, n.1-2, p.275-290, 2007.

SUPPADIT, T.; PHUMKOKRAK, N.; POUNGSUK, P. The effect of using quail litter biochar on soybean (*Glycine max* [L.] Merr.) production. Chilean Journal of Agricultural Research, v.72, n.2, 2012.

TAGOE, S.O.; HORIUCHI, T.; MATSUI, T. Preliminary evaluation of the effects of carbonized chicken manure, refuse derived fuel and K fertilizer application on the growth, nodulation, yield, N and P contents of soybean and cowpea in the greenhouse. African Journal of Agricultural Research, v.3, n.11, p.759-774, 2008.

TAIZ, L.; ZEIGER, E. Plant physiology. 5 ed. Porto Alegre: Artmed, 2013. 918p.

TESTER, M.; DAVENPORT, R. $^{Na+}$ tolerance and $^{Na+}$ transport in higher plants. Annals of Botany, v.91, n.5, p.503-527, 2003.

THOMAS, S.C.; FRYE, S.; GALE, N.; GARMON, M.; LAUNCHBURY, R.; MACHADO, N.; MELAMED, S.; MURRAY, J.; PETROFF, A.; WINSBOROUGHT, C. Biochar mitigates negative effects of salt additions on two herbaceous plant species. Journal of Environmental Management, v.129, s.n., p.62-68, 2013.

U.N. - United Nations. Population Division World Population Prospects: The 2012 Revision Population.<http://esa.un.org/unpd/wpp/index.htm>. 25 Out. 2013.

UCHÔA, S.C.P.; IVANOFF, M.E.A.; ALVES, J.M.A.; SEDIYAMA, T.; MARTINS, S.A. Potassium fertilization in cover on production components of sunflower cultivars. Revista Ciência Agronômica, v.42, n.1, p.8-15, 2011.

UNGARO, M.R.G. Cultura do girassol. Campinas: IAC, 2000. 36 p.

USDA - United States Department of Agriculture. <http://www.usda.gov/wps/portal/usda/usdahome>. 09 Set. 2014.

UTOMO, W.H.; GURITNO, B.; SOEHONO, L.A. The effect of biochar on the growth and N fertilizer requirement of maize (*Zea mays* L.) in green house experiment. Journal of Agricultural Science, v.4, n.5, p.255-262, 2012.

WANG, J.; ZHANG, M.; XIONG, Z.; LIU, P.; PAN, G. Effects of biochar addition on N2O and $_{CO2}$ emissions from two paddy soils. Biology and Fertility of Soils, v.47, n.8, p.887-896, 2011.

WATANABE, A.A. Desenvolvimento de plantas de girassol (*Helianthus annuus* L. cv. Pacino) com variação de nutrientes na solução nutritiva e aplicação de aminozide. Botucatu: UNESP, 2007. 106p. Master's thesis.

WENDT, V.; BÜLL, L.T.; CORRÊA, J.C.; CRUSCIOL, C.A.C. Sunflower production in two sowing systems as a function of winter green manure associated with NPK doses. Acta Scientiarum Agronomy, v.27, n.4, p. 617-621, 2005.

WOOLF, D.; AMONETTE, J. E.; STREE-PERROTT, F. A.; LEHMANN, J.; JOSEPH S. Sustainable biochar to mitigate global climate change. Nature Communications, v.1, n.56, p.1-8, 2010.

YANAI, Y.; TOYOTA, K.; OKAZAKI, M. Effects of charcoal addition on N2O emissions from soil resulting from rewetting air-dried soil in short-term laboratory experiments. Japanese Society of Soil Science and Plant Nutrition, v. 53, n.2, p.181-188, 2007.

YASIN, M.; MAHMOOD, A.; ALI,A.; AZIZ1,M.; JAVAID, M.M.; IQBAL, Z.; TANVEER, A. Impact of varying planting patterns and fertilizer application strategies on autumn planted sunflower hybrid. Cercetãrí Agronomice în Moldova, v.46, n.2, p.39-51, 2013.

ZAGONEL, J.; MUNDSTOCK, C.M. Doses and epocas de aplicação de nitrogen em cobertura em duas cultivares de girassol. Pesquisa Agropecuária Brasileira, v.26, n.9, p.1487-1492, 1991.

ZOBIOLE, L.H.S., CASTRO, C., OLIVEIRA, F.A., OLIVEIRA JUNIOR, A. Marcha de absorção de macronutrientes na cultura do girassol. Revista Brasileira de Ciências do Solo, v.34, n.2, p.425-433, 2010.

Printed by Books on Demand GmbH, Norderstedt / Germany